ALSO BY CARL ZIMMER

At the Water's Edge:
Fish with Fingers, Whales with Legs, and How Life Came Ashore but Then Went Back to the Sea

Parasite Rex:
Inside the Bizarre World of Nature's Most Dangerous Creatures

Evolution: The Triumph of an Idea

Soul Made Flesh:
The Discovery of the Brain and How It Changed the World

Smithsonian Intimate Guide to Human Origins

MICROCOSM

MICROCOSM

E. coli and the New Science of Life

CARL ZIMMER

Pantheon Books, New York

All illustrations by Tadeusz Majewski unless otherwise noted.

Library of Congress Cataloging-in-Publication Data
Zimmer, Carl, [date]
Microcosm : E. coli and the new science of life / Carl Zimmer.
p. cm.
Includes bibliographical references and index.
ISBN 978-0-375-42430-4
1. Escherichia coli. 2. Microbiology—History. 3. Molecular biology—History. 4. Genetics—History. I. Title. [DNLM: 1. Escherichia coli. 2. Microbiology—history. 3. Genetics—history. 4. History, 20th century. 5. Molecular biology—history. QW 11.1 Z72m 2008]
QR82.E6Z56 2008 579.3'42—dc22 2007037155

www.pantheonbooks.com

Printed in the United States of America

First Edition

2 4 6 8 9 7 5 3 1

TO VERONICA,

OUR LATEST LOVELY LIFE

CONTENTS

MICROCOSM

One

SIGNATURE

I GAZE OUT A WINDOW, a clear, puck-shaped box in my hand. Life fills my view: fescue and clover spreading out across the yard, rose of Sharon holding out leaves to catch sunlight and flowers to lure bumblebees. An orange cat lurks under a lilac bush, gazing up at an oblivious goldfinch. Snowy egrets and seagulls fly overhead. Stinkhorns and toadstools rudely surprise. All of these things have something in common with one another, something not found in rocks or rivers, in tugboats or thumbtacks. They live.

The fact that they live may be obvious, but what it means for them to be alive is not. How do all of the molecules in a snowy egret work together to keep it alive? That's a good question, made all the better by the fact that scientists have decoded only a few snips of snowy egret DNA. Most other species on Earth are equally mysterious. We don't even know all that much about ourselves. We can now read the entire human genome, all 3.5 billion base pairs of DNA in which the recipe for *Homo sapiens* is written. Within this genetic tome, scientists have identified about 18,000 genes, each of which encodes proteins that build our bodies. And yet scientists have no idea what a third of those genes are for and only a faint understanding of most of the others. Our ignorance actually reaches far beyond protein-coding genes. They take up only about 2 percent of the human genome. The other 98 percent of our DNA is a barely explored wilderness.

Only a few species on the entire planet are exceptions to this rule. The biggest exception lives in the plastic box in my hand. The box—a petri dish—looks lifeless compared with the biological riot outside my window. A few beads of water cling to the underside of the lid. On the bottom is a layer of agar, a firm gray goo made from dead algae and infused with sugar and other compounds. On top of the agar lies a trail of pale gold spots, a pointillistic flourish. Each of those spots is made up of millions of bacteria. They belong to a species that scientists have studied intensely for a century, that they understand better than almost any other species on the planet. I've made this species my guide—an oracle that can speak of the difference between life and lifeless matter, of the rules that govern all living things: bacteria, snowy egret, and curious human. I turn over the dish. On the bottom is a piece of tape labeled "*E. coli* K-12 (P1 strain)."

I got my dish of *Escherichia coli* on a visit to Osborne Memorial Laboratories, a fortress of a building on the campus of Yale University. On the third floor is a laboratory filled with nose-turning incubators and murky flasks. A graduate student named Nadia Morales put on purple gloves and set two petri dishes on a lab bench. One was sterile, and the other contained a cloudy mush rich with *E. coli.* She picked up a loop—a curled wire on a plastic handle—and stuck it in the flame of a Bunsen burner. The loop glowed orange. She moved it away from the flame, and after it cooled down she dipped it into the mush. Opening the empty dish, Morales smeared a dollop across the sterile agar as if she were signing it. She snapped the lid on the second dish and taped it shut.

"You'll probably start seeing colonies tomorrow," she said, handing it to me. "In a few days it will get stinky."

It was as if Morales had given me the philosopher's stone. The lifeless agar in my petri dish began to rage with new chemistry. Old molecules snapped apart and were forged together into new ones. Oxygen molecules disappeared from the air in the dish, and carbon dioxide and beads of water were created. Life had taken hold. If I had microscopes for eyes, I could have watched the hundreds of *E. coli* Morales had given me as they wandered, fed, and grew. Each one is shaped like a microscopic submarine, enshrouded by fatty, sugary membranes. It trails propeller-like tails that spin hundreds of times a second. It is packed with tens of millions of molecules, jostling and cooperating to make the microbe grow. Once it grows long enough, it splits cleanly in two. Splitting again and again, it gives rise to a miniature dynasty. When these dynasties grow large enough, they become visible as golden spots. And together the spots reveal the path of Morales's living signature.

E. coli may seem like an odd choice as a guide to life if the only place you've heard about it is in news reports of food poisoning. There are certainly some deadly strains in its ranks. But most *E. coli* are harmless. Billions of them live peacefully in my intestines, billions more in yours, and many others in just about every warm-blooded animal on Earth. All told, there are around 100 billion billion *E. coli* on Earth. They live in rivers and lakes, forests and backyards. And they also live in thousands of laboratories, nurtured in yeasty flasks and smeared across petri dishes.

In the early twentieth century, scientists began to study harmless strains of *E. coli* to understand the nature of life. Some of them marched to Stockholm in the late 1900s to pick up Nobel Prizes for their work. Later generations of scientists probed even further into *E. coli*'s existence, carefully studying most of its 4,000-odd genes and discovering more rules to life. In *E. coli,* we can begin to see how genes must work together to sustain life, how life can defy the universe's penchant for disorder and chaos. As a single-celled microbe, *E. coli* may not seem to have much in common with a complicated species like our own. But scientists keep finding more parallels between its life and ours. Like us, *E. coli* must live alongside other members of its species, in cooperation, conflict, and conversation. And like us, *E. coli* is the product of evolution. Scientists can now observe *E. coli* as it evolves, mutation by mutation. And in *E. coli,* scientists can see an ancient history we also share, a history that includes the origin of complex features in cells, the common ancestor of all living things, a world before DNA. *E. coli* can not only tell us about our own deep history but can also reveal things about the evolutionary pressures that shape some of the most important features of our existence today, from altruism to death.

Through *E. coli* we can see the history of life, and we can see its future as well. In the 1970s, scientists first began to engineer living things, and the things they chose were *E. coli.* Today they are manipulating *E. coli* in even more drastic ways, stretching the boundaries of what we call life. With the knowledge gained from *E. coli,* genetic engineers now transform corn, pigs, and fish. It may not be long before they set to work on humans. *E. coli* led the way.

I hold the petri dish up to the window. I can see the trees and flowers through its agar gauze. Each spot of the golden signature refracts their image. I look at life through a lens made of *E. coli.*

Two

E. COLI AND THE ELEPHANT

"LUXURIOUS GROWTH"

***ESCHERICHIA COLI* HAS LURKED WITHIN** our ancestors for millions of years, before our ancestors were even human. It was not until 1885 that our species was formally introduced to its lodger. A German pediatrician named Theodor Escherich was isolating bacteria from the diapers of healthy babies when he noticed a rod-shaped microbe that could produce, in his words, a "massive, luxurious growth." It thrived on all manner of food—milk, potatoes, blood.

Working at the dawn of modern biology, Escherich could say little more about his new microbe. What took place within *E. coli*—the transformation of milk, potatoes, or blood into living matter—was mostly a mystery in the 1880s. Organisms were like biological furnaces, scientists agreed, burning food as fuel and creating heat, waste, and organic molecules. But they debated whether this transformation required a mysterious vital spark or was just a variation on the chemistry they could carry out themselves in their laboratories.

Bacteria were particularly mysterious in Escherich's day. They seemed fundamentally different from animals and other forms of multicellular life. A human cell, for example, is thousands of times larger than *E. coli.* It has a complicated inner geography dominated by a large sac known as the nucleus, inside of which are giant structures called chromosomes. In bacteria, on the other hand, scientists could find no nucleus, nor much of anything else. Bacteria seemed like tiny, featureless bags of goo that hovered at the boundary of life and nonlife.

Escherich, a forward-thinking pediatrician, accepted a radical new theory about bacteria: far from being passive goo, they infected people and caused diseases. As a pediatrician, Escherich was most concerned with diarrhea, which he called "this most murderous of all intestinal disease." A horrifying number of infants died of diarrhea in nineteenth-century Germany, and doctors did not understand why. Escherich was convinced—rightly—that bacteria were killing the babies. It would be no simple matter to find those pathogens, however, because the guts of the healthiest babies were rife with bacteria. Escherich would have to sort out the harmless species of microbes before he could recognize the killers.

"It would appear to be a pointless and doubtful exercise to examine and disentangle the apparently randomly appearing bacteria," he wrote. But he tried anyway, and in that survey he came across a harmless-seeming resident we now call *E. coli.*

Escherich published a brief description of *E. coli* in a German medical journal, along with a little group portrait of rod-shaped microbes. His discovery earned no headlines. It was not etched on his gravestone when he died, in 1911. *E. coli* was merely one of a rapidly growing list of species of bacteria that scientists were discovering. Yet it would become Escherich's great legacy to science.

Its massive, luxurious growth would bloom in laboratories around the world. Scientists would run thousands of experiments to understand its growth—and thereby to understand the fundamental workings of life. Other species would also do their part in the rise of modern biology. Flies, watercress, vinegar worms, and bread mold all had their secrets to share. But the story of *E. coli* and the story of modern biology are extraordinarily intertwined. When scientists were at loggerheads over some basic question of life—what are genes made of? do all living things have genes?—it was often *E. coli* that served as the expert witness. By understanding how *E. coli* produced its luxurious growth—how it survived, fed, and reproduced—biologists went a great way toward understanding the workings of life itself. In 1969, when the biologist Max Delbrück accepted a Nobel Prize for his work on *E. coli* and its viruses, he declared, "We may say in plain words, 'This riddle of life has been solved.' "

THE UNITY OF LIFE

Escherich originally dubbed his bacteria *Bacterium coli communis:* a common bacterium of the colon. In 1918, seven years after Escherich's death, scientists renamed it in his honor. By the time it got a new name, it had taken on a new life. Microbiologists were beginning to rear it by the billions in their laboratories.

In the early 1900s, many scientists were pulling cells apart to see what they were made of, to figure out how they turned raw material into living matter. Some scientists studied cells from cow muscles, others sperm from salmon. Many studied bacteria, including *E. coli.* In all of the living things they dissected, scientists discovered the same basic collection of molecules. They focused much of their attention on proteins. Some proteins give life its structure—the collagen in skin, the keratin in a horse's hoof. Other proteins, known as enzymes, usher other molecules into chemical reactions. Some enzymes split atoms off molecules, and others weld molecules together.

Proteins come in a maddening diversity of complicated shapes, but scientists discovered that they also share an underlying unity. Whether from humans or bacteria, proteins are all made from the same building blocks: twenty small molecules known as amino acids. And these proteins work in bacteria much as they do in humans. Scientists were surprised to find that the same series of enzymes often carry out the same chemical reactions in every species.

"From the elephant to butyric acid bacterium—it is all the same!" the Dutch biochemist Albert Jan Kluyver declared in 1926.

The biochemistry of life might be the same, but for scientists in the early 1900s, huge differences seemed to remain. The biggest of all was heredity. In the early 1900s, geneticists began to uncover the laws by which animals, plants, and fungi pass down their genes to their offspring. But bacteria such as *E. coli* didn't seem to play by the same rules. They did not even seem to have genes at all.

Much of what geneticists knew about heredity came from a laboratory filled with flies and rotten bananas. Thomas Hunt Morgan, a biologist at Columbia University, bred the fly *Drosophila melanogaster* to see how the traits of parents are passed on to their offspring. Morgan called the factors that control the traits genes, although he had no idea what genes actually were. He did know that mothers and fathers both contributed copies of genes to their offspring and that sometimes a gene could fail to produce a trait in one generation only to make it in the next. He could breed a red-eyed fly with a white-eyed one and get a new generation of flies with only red eyes. But if he bred those hybrid flies with each other, the eyes of some of the grandchildren were white.

Morgan and his students searched for molecules in the cells of *Drosophila* that might have something to do with genes. They settled on the fly's chromosomes, those strange structures inside the nucleus. When chromosomes are given a special stain, they look like crumpled striped socks. The stripes on *Drosophila* chromosomes, Morgan and his students discovered, are as distinctive as bar codes. Chromosomes mostly come in pairs, one inherited from each parent. And by comparing their stripes, Morgan and his students demonstrated that chromosomes can change from one generation to the next. As a fly's sex cells develop, each pair of chromosomes embrace and swap segments. The segments a fly inherited determined which genes it carried.

There was something almost mathematically abstract about these findings. George Beadle, one of Morgan's graduate students, decided to bring genes down to earth by figuring out exactly how they controlled a single trait, such as eye color. Working with the biochemist Edward Tatum, Beadle tried to trace cause and effect from a fly's genes to the molecules that make up the pigment in its eyes. But that experiment soon proved miserably complex. Beadle and Tatum abandoned flies for a simpler species: the bread mold *Neurospora crassa.*

Bread mold may not have obvious traits such as eyes and wings, but it does produce many enzymes, some of which build amino acids. To see how the mold's genes control those enzymes, Beadle and Tatum bombarded it with X-rays. They knew that when fly larvae are exposed to X-rays, the radiation mutates some of their genes. The mutations produce new traits—extra leg bristles or a different eye color—which mutant flies can pass down to their offspring.

Beadle and Tatum now created bread mold mutants. Some were unable to produce certain types of amino acids because they now lacked a key enzyme. But if Beadle and Tatum mated the mutant bread mold with a normal one, some of their offspring could make the amino acid once more. Beadle and Tatum concluded in 1941 that behind each enzyme in bread mold there is one gene.

A hazy but consistent picture of genes was emerging—at least a picture of the genes of animals, plants, and fungi. But there didn't seem to be a place for bacteria in the picture. The best evidence for genes came from chromosomes, and bacteria seemed to have no chromosomes at all. Even if bacteria did have genes, scientists had little hope of finding them. Scientists could study a fly's genes thanks to the fact that flies reproduce sexually. A fly's chromosomes get cut up and shuffled in different combinations in its offspring. Scientists could not run this sort of experiment on bacteria, because bacteria did not have sex. They seemed to just grow and then split in two. Many researchers looked at bacteria as simply loose bags of enzymes—a fundamentally different kind of life.

It would turn out, however, that all life, bacteria included, shares the same foundation. *E. coli* would reveal much of that unity, and in the process it would become one of the most powerful tools biologists could use to understand life.

The transformation started with a simple question. Edward Tatum wondered if the one-gene, one-enzyme rule he discovered in mold applied to bacteria. He decided to run the mold experiment again, this time directing his X-rays at bacteria. For his experiment, Tatum chose a strain of *E. coli* called K-12. It had been isolated in 1922 from a California man who suffered from diphtheria, and it had been kept alive ever since at Stanford University, where it was used for microbiology classes.

Tatum's choice was practical. Like most strains of *E. coli,* K-12 is harmless. *E. coli* is also versatile enough to build all of its own amino acids and many other molecules. For food, it needs little more than sugar, ammonia, and some trace minerals. If *E. coli* used a lot of enzymes to turn this food into living matter, Tatum would have plenty of targets for his X-rays. He might succeed in creating only a few mutants of the sort he was looking for, but thanks to *E. coli*'s luxurious growth he'd be able to see them. A single mutant could give rise to a visible colony in a day.

Tatum pelted colonies of *E. coli* with enough X-rays to kill 9,999 of every 10,000 bacteria. Among the few survivors he discovered mutants that could grow only if he supplied them with a particular amino acid. Helped along, the mutants could even reproduce, and their offspring were just as crippled. Tatum had gotten the same results as he had with bread mold. It looked as if behind every enzyme in *E. coli* lurked a gene.

It was a profound discovery, but Tatum remained cautious about its significance. It now seemed that bacteria had genes, but he could not say for sure. The best way to prove that a species had genes was to breed males and females and study their offspring. But *E. coli* seemed sadly celibate.

"The term 'gene' can therefore be used in connection with bacteria only in a general sense," Tatum wrote.

The connection became far stronger when a somber young student arrived at Tatum's lab at Yale. Joshua Lederberg was only twenty-one years old when he began to work with Tatum, but he had a grand ambition: to find out whether bacteria had sex. As part of his military service during World War II, Lederberg had spent time in a naval hospital on Long Island, where he examined malaria parasites from marines fighting in the Pacific. He had gazed down at the single-celled protozoans, which sometimes reproduced by dividing and sometimes by taking male and female forms and mating. Perhaps bacteria had this sort of occasional sex, and no one had noticed. Others might mock the idea as a fantasy, but Lederberg decided to take what he later called "the long-shot gamble in looking for bacterial sex."

When Lederberg heard about Tatum's work, he realized he could look for bacterial sex with a variation on Tatum's experiments. Tatum was amassing a collection of mutant *E. coli* K-12, including double mutants—bacteria that had to be fed two compounds to survive. Lederberg reasoned that if he mixed two different double mutants together, they might be able to pick up working versions of their genes through sex.

Lederberg started work at Yale in 1946. He selected a mutant strain that could make neither the amino acid methionine nor biotin, a B vitamin. The other strain he picked couldn't make the amino acids threonine and proline. Lederberg put the bacteria in a broth he stocked with all four compounds so that the mutant microbes could grow and multiply. They mingled in the broth for a few weeks, with plenty of opportunity for hypothetical sex.

Lederberg drew out samples of the bacteria and put them on fresh petri dishes. Now he withheld the four nutrients they could not make themselves: threonine, proline, methionine, and biotin. Neither of the original mutant strains could grow in the dishes. If their descendants were simply copies of their ancestors, Lederberg reasoned, they would stop growing as well.

But after weeks of frustration—of ruined plates, of dead colonies—Lederberg finally saw *E. coli* spreading across his dishes. A few microbes had acquired the ability to make all four amino acids. Lederberg concluded that their ancestors must have combined their genes in something

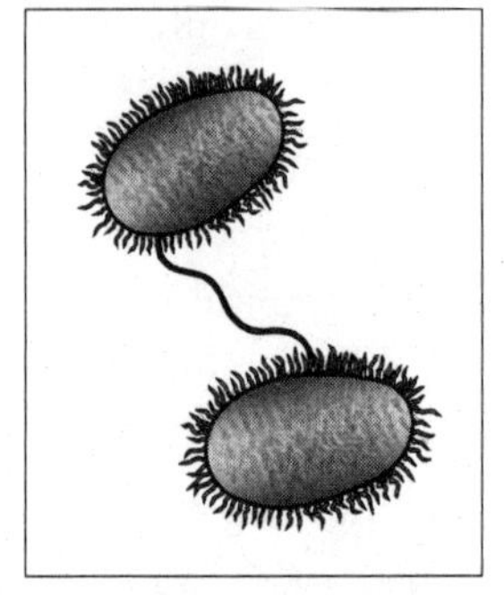

Two *E. coli* having bacterial sex

akin to sex. And in their sex they proved that they carried genes.

In the years that followed, the discovery would allow scientists to breed *E. coli* like flies and to probe genes far more intimately than ever before. Twelve years later, at the ancient age of thirty-three, Lederberg would share the Nobel Prize in Medicine with Tatum and Beadle. But in 1946, when he picked up his petri dishes and noticed the spots that appeared to be the sexual colonies he had dreamed of, Lederberg allowed himself just a single word alongside the results in his notebook: "*Hooray.*"

HOST AND PARASITE

While Lederberg was observing *E. coli* having sex, other scientists were observing it getting sick. And they were learning things that were just as important about the nature of life.

The first scientist to appreciate just how revealing a sick *E. coli* could be was not a biologist but a physicist. Max Delbrück had originally studied under Niels Bohr and the other pioneers of quantum physics. In the 1930s it seemed as if a few graceful equations could melt away many of the great mysteries of the universe. But life would not submit. Physicists like Delbrück were baffled by life's ability to store away all of the genes necessary to build a kangaroo or a liverwort in a single cell. Delbrück decided to make life—and in particular, life's genes—his study.

"The gene," Delbrück proposed, "is a polymer that arises by the repetition of identical atomic structures." To discover the laws of that polymer, he came to the United States, joining Morgan's laboratory to breed flies. But the physicist in Delbrück despised the messy quirks of *Drosophila*. He craved another system that could provide him with far more data and was far simpler. As luck would have it, another member of Morgan's lab, Emory Ellis, was studying the perfect one: the viruses that infect *E. coli.*

The viruses that infect *E. coli* were too small for Delbrück and Ellis to see. As best anyone could tell, they infected their bacterial hosts and reproduced inside, killing the microbes and wandering off to find new victims. The new viruses seemed identical to the old, which suggested that they might carry genes. Delbrück and Ellis set out to chart the natural history of *E. coli*'s viruses.

To study the viruses—known as bacteriophages—Delbrück and Ellis could look only for indirect clues. If they added viruses to a dish of *E. coli,* the viruses invaded the bacteria and replicated inside them. The new viruses left behind the shattered remains of their hosts and infected new ones. Over a few hours spots formed on the dish where their victims formed transparent pools of carnage. "Bacterial viruses make themselves known by the bacteria they destroy," Delbrück said, "as a small boy announces his presence when a piece of cake disappears."

Although the signs of the viruses were indirect, there were a lot of them. Billions of new viruses could appear in a dish in a few hours. The power of Delbrück and Ellis's system attracted a small flock of young scientists. They called themselves the Phage Church, and Delbrück was their pope. The Phage Church demonstrated that *E. coli*'s bacteriophages were not all alike. Some could infect certain *E. coli* strains but not others. By triggering mutations in the viruses, the scientists could cause the viruses to infect new strains. The ability to infect *E. coli* passed down from virus to virus. Viruses, it became clear, had genes—genes that must be very much like those of their host, *E. coli.*

The genes of host and parasite are so similar, in fact, that scientists discovered certain kinds of viruses that could merge into *E. coli,* blurring their identities. These prophages, as they are called, can invade *E. coli* and then disappear. A prophage's hosts behave normally, growing and dividing like their virus-free neighbors. Yet scientists found that the prophages survived within *E. coli,* which passed them down from one generation to the next. To rouse a prophage, the scientists needed only to expose a dish of infected *E. coli* to a flash of ultraviolet light. The bacteria abruptly burst open with hundreds of new prophages, which began to infect new hosts, leaving behind the clear pools of destruction. Two had become one, only to become two again.

THE STUFF OF GENES

In the merging dance of *E. coli* and its viruses, the Phage Church discovered clues to some of life's great questions. And for them there was no greater question than what genes are made of.

Until the 1950s, most scientists suspected that proteins were the stuff of genes. They had no direct evidence but many powerful hints. Genes exist in all living things, even bacteria and viruses, and proteins appeared to be in all of them as well. Scientists studying flies had located genes in the chromosomes, and chromosomes contain proteins. Scientists also assumed that the molecules from which genes are made had to be complicated, since genes somehow gave rise to all the complexity of life. Proteins, scientists knew, often are staggeringly intricate. All that remained was to figure out how proteins actually function as genes.

The first major challenge to this vague consensus came in 1944, when a physician announced that genes are not in fact made of protein. Oswald Avery, who worked at the Rockefeller Institute in New York, studied the bacteria *Pneumococcus.* It comes in both a harmless form and a dangerous one that can cause pneumonia. Earlier experiments had hinted that genes control the behaviors of the different strains. If scientists killed the dangerous strain before injecting it into mice, it did not make the mice sick. But if the dead strain was mixed with living harmless *Pneumococcus,* an injection killed the mice. The harmless strain had been transformed into pathogens, and their descendants remained deadly. In other words, genetic material had moved from the dead strain to the live one.

Avery and his colleagues isolated compound after compound from the deadly strain and added each one to the harmless strain. Only one molecule, they found, could make the harmless strain deadly. It was not a protein. It was something called deoxyribonucleic acid, DNA for short.

Scientists had known of DNA for decades but didn't know what to make of it. In 1869, a Swiss biochemist named Johann Miescher had discovered a phosphorus-rich goo in the pus on the bandages of wounded soldiers. The goo came to be known as nucleic acid, which scientists later discovered comes in two nearly identical forms: ribonucleic acid (RNA) and deoxyribonucleic acid. The phosphorus in DNA helps form a backbone, along with oxygen and sugar. Connected to this backbone are four kinds of compounds, known as bases, rich in carbon and nitrogen.

DNA was clearly important to life, because scientists could find it in just about every kind of cell they looked at. It could even be found in fly chromosomes, where genes were known to reside. But many researchers thought DNA simply offered some kind of physical support for chromosomes—it might wind around genes like cuffs. Few thought DNA had enough complexity to be the material of genes. DNA was, as Delbrück once put it, "so *stupid* a substance."

Stupid or not, DNA is what genes are made of, Avery concluded. But his experiments failed to win over hardened skeptics, who wondered if his purified DNA had actually been contaminated by some proteins.

It would take another decade of research on *E. coli* and its viruses to start to redeem DNA's reputation. While Avery was sifting *Pneumococcus* for genes, Delbrück's Phage Church was learning how to see *E. coli*'s viruses. The viruses were no longer mathematical abstractions but hard little creatures. Using the newly invented electron microscope, Delbrück and his colleagues discovered that bacteriophages are elegantly geometrical shells. After a phage lands on *E. coli,* it sticks a needle into the microbe and injects something into its new host. The shell remains sitting on *E. coli*'s surface, an empty husk, while the virus's genes enter the microbe.

The life cycle of *E. coli*'s viruses opened up the chance to run an elegantly simple experiment. Alfred Hershey and Martha Chase, two scientists at the Cold Spring Harbor Laboratory on Long Island, created viruses with radioactive tracers in their DNA. They allowed the viruses to infect *E. coli* and then pulled off their empty husks in a fast-spinning centrifuge. Hershey and Avery searched for radioactivity and found it only within the bacteria, not the virus shells.

Hershey and Chase then reversed the experiment, spiking the protein in the viruses with radioactive tracers. Once the viruses had infected *E. coli,* only the empty shells were radioactive. A

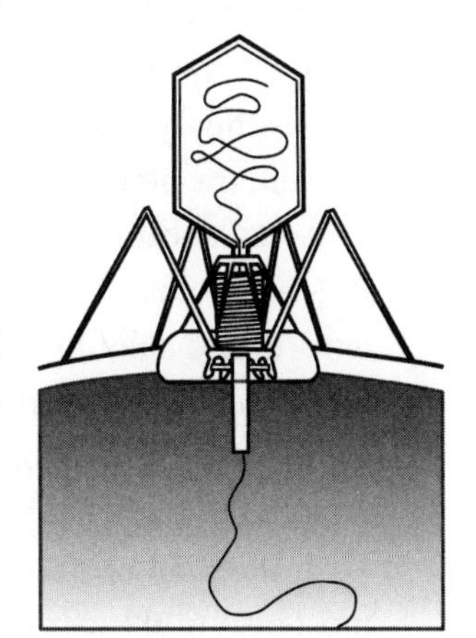

A virus inserts its DNA into *E. coli.*

decade after Avery's experiment, Hershey and Chase confirmed his conclusion: genes are made of DNA.

No one was more excited by the new results than a young American biologist named James Watson. Watson was only twenty when he was initiated into the Phage Church, blasting *E. coli*'s viruses with X-rays for his dissertation work. He was taught the conventional view that genes are made of proteins, but his own research was drawing his attention to DNA. He saw Hershey and Chase's experiment as "a powerful new proof that DNA is the primary genetic material."

In order to understand how DNA acts as genetic material, however, it was necessary to figure out its structure. Watson was working at the time at the University of Cambridge, where he quickly teamed up with Francis Crick, a British physicist who also wanted to understand the secret of life. Together they pored over clues about DNA and tinkered with arrangements of phosphates, sugars, and bases. In February 1953, they suddenly figured out its shape. They assembled a towering model of steel plates and rods. It was a twisted ladder of sugar and phosphates, with bases for rungs.

The structure was beautiful, simple, and eloquent. It seemed to practically speak for itself about how genes work. Each phosphate strand is studded with billions of bases, arrayed in a line like a string of text. The text can have an infinite number of meanings, depending on how the bases are arranged. By this means, DNA stores the information necessary for building any protein in any species.

The structure of DNA also suggested to Watson and Crick how it could be reproduced. They envisioned the strands being pulled apart, and a new strand being added to each. Building a new DNA strand would be simplified by the fact that each kind of base can bond to only one other kind. As a result, the new strands would be perfect counterparts.

It was a beautiful idea, but it didn't have much hard evidence going for it. Max Delbrück worried about what he called "the untwiddling problem." Could a double helix be teased apart and transformed into two new DNA molecules without creating a tangled mess? Delbrück tried to answer the question but failed. Success finally came in 1957, to a graduate student and a postdoc at Caltech, Matthew Meselson and Frank Stahl. With the help of *E. coli,* they conducted what came to be known as the most beautiful experiment in biology.

Meselson and Stahl realized that they could trace the replication of

DNA by raising *E. coli* on a special diet. *E. coli* needs nitrogen to grow, because the element is part of every base of DNA. Normal nitrogen contains fourteen protons and fourteen neutrons, but lighter and heavier forms of nitrogen also exist, with fewer or more neutrons. Meselson and Stahl fed *E. coli* ammonia laced with heavy nitrogen in which each atom carried a fifteenth neutron. After the bacteria had reproduced for many generations, they extracted some DNA and spun it in a centrifuge. By measuring how far the DNA moved as it was spun, they could calculate its weight. They could see that the DNA from *E. coli* raised on heavy nitrogen was, as they had expected, heavier than DNA from normal *E. coli.*

Meselson and Stahl then ran a second version of the experiment. They moved some of the heavy-nitrogen *E. coli* into a flask where they could feed on normal nitrogen, with only fourteen neutrons apiece. The bacteria had just enough time to divide once before Meselson and Stahl tossed their DNA in the centrifuge. If Watson and Crick were right about how DNA reproduced, Meselson and Stahl knew what to expect. Inside each microbe, the heavy strands would have been pulled apart, and new strands made from light nitrogen would have been added to them. The DNA in the new generation of *E. coli* would be half heavy, half light. It should form a band halfway between where the light and heavy forms did. And that was precisely what Meselson and Stahl saw.

Watson and Crick might have built a beautiful model. But it took a beautiful experiment on *E. coli* for other scientists to believe it was also true.

A UNIVERSAL CODE

The discovery of *E. coli*'s sex life gave scientists a way to dissect a chromosome. It turned out that *E. coli* has a peculiar sort of sex, with one microbe casting out a kind of molecular grappling hook to reel in a partner. Its DNA moves into the other microbe over the course of an hour and a half. Élie Wollman and François Jacob, both at the Pasteur Institute in Paris, realized that they could break off this liaison. They mixed mutants together and let them mate for a short time before throwing them into a blender. Depending on how long the bacteria were allowed to mate, the recipient might or might not get a gene it needed to survive. By timing how long it took various genes to enter *E. coli*, Wollman and Jacob could create a genetic map. It turned out that *E. coli*'s genes are arrayed on a chromosome shaped in a circle.

Scientists also discovered that along with its main chromosome *E. coli* carries extra ringlets of DNA, called plasmids. Plasmids carry genes of their own, some of which they use to replicate themselves. Some plasmids also carry genes that allow them to move from one microbe to another. *E. coli* K-12's grappling hooks, for example, are encoded by genes on plasmids. Once the microbes are joined, a copy of the plasmid's DNA is exchanged, along with some of the chromosome itself.

As some scientists mapped *E. coli*'s genes, others tried to figure out how their codes are turned into proteins. At the Carnegie Institution in Washington, D.C., researchers fed *E. coli* radioactive amino acids, the building blocks of proteins. The amino acids ended up clustered around pellet-shaped structures scattered around the microbe, known as ribosomes. Loose amino acids went into the ribosomes, and full-fledged proteins came out. Somehow the instructions from *E. coli*'s DNA had to get to the ribosomes to tell them what proteins to make.

It turned out that *E. coli* makes special messenger molecules for the job. The first step in making a protein requires an enzyme to clamp on to a gene and crawl along its length. It builds a single-stranded version of the gene from RNA. This RNA can then move to a ribosome, delivering its genetic message.

How a ribosome reads that message was far from clear, though. RNA, like DNA, is made of four different bases. Proteins are combinations of twenty amino acids. *E. coli* needs some kind of dictionary to translate instructions written in the language of genes into the language of proteins.

In 1957, Francis Crick drafted what he imagined the dictionary might look like. Each amino acid was encoded by a string of three bases, known as a codon. Marshall Nirenberg and Heinrich Matthaei, two scientists at the National Institutes of Health, soon began an experiment to see if Crick's dictionary was accurate. They ground up *E. coli* with a mortar and pestle and poured its innards into a series of test tubes. To each test tube they added a different type of amino acid. Then Nirenberg and Matthaei added the same codon to each tube: three copies of uracil (a base found in RNA but not in DNA). They waited to see if the codon would recognize one of the amino acids.

In nineteen tubes nothing happened. The twentieth tube was filled with the amino acid phenylalanine, and only in that tube did new proteins form. Nirenberg and Matthaei had discovered the first entry in life's dictionary: UUU equals phenylalanine. Over the next few years they and other scientists would decipher *E. coli*'s entire genetic code.

Having deciphered the genetic code of a species for the first time, Nirenberg and his colleagues then compared *E. coli* to animals. They filled test tubes with the crushed cells of frogs and guinea pigs, and added codons of RNA to them. Both frogs and guinea pigs followed the same recipe for building proteins as *E. coli* had. In 1967, Nirenberg and his colleagues announced they had found "an essentially universal code."

Nirenberg would share a Nobel Prize for Medicine the following year. Delbrück got his the year after. Lederberg, Tatum, and many others who worked on *E. coli* were also summoned to Stockholm. A humble resident of the gut had led them to glory and to a new kind of science, known as molecular biology, that unified all of life. Jacques Monod, another of *E. coli*'s Nobelists, gave Albert Kluyver's old claim a new twist, one that many scientists still repeat today.

"What is true for *E. coli* is true for the elephant."

THE SHAPE OF LIFE

With the birth of molecular biology, genes came to define what it means to be alive. In 2000, President Bill Clinton announced that scientists had completed a rough draft of the human genome—the entire sequence of humans' DNA. He declared, "Today, we are learning the language in which God created life."

But on their own, genes are dead, their instructions meaningless. If you coax the chromosome out of *E. coli*, it cannot build proteins by itself. It will not feed. It will not reproduce. The fragile loop of DNA will simply fall apart. Understanding an organism's genes is only the first step in understanding what it means for the organism to be alive.

Many biologists have spent their careers understanding what it means for *E. coli* in particular to be alive. Rather than starting from scratch with another species, they have built on the work of earlier generations. Success has bred more success. In 1997, scientists published a map of *E. coli*'s K-12's entire genome, including the location of 4,288 genes. The collective knowledge about *E. coli* makes it relatively simple for a scientist to create a mutant missing any one of those genes and then to learn from its behavior what that gene is for. Scientists now have a good idea of what all but about 600 genes in *E. coli* are for. From the hundreds of thousands of papers scientists have published on *E. coli* comes a portrait of a living thing governed by rules that often apply, in one form or another, to all life. When Jacques Monod boasted of *E. coli* and the elephant, he was speaking only of genes and proteins. But *E. coli* turns out to be far more complex—and far more like us—than Monod's generation of scientists realized.

The most obvious thing one notices about *E. coli* is that one can notice *E. coli* at all. It is not a hazy cloud of molecules. It is a densely stuffed package with an inside and an outside. Life's boundaries take many forms. Humans are wrapped in soft skin, crabs in a hard exoskeleton. Redwoods grow bark, squid a rubbery sheet. *E. coli*'s boundary is just a few hundred atoms thick, but it is by no means simple. It is actually a series of layers within layers, each with its own subtle structure and complicated jobs to carry out.

E. coli's outermost layer is a capsule of sugar teased like threads of cotton candy. Scientists suspect it serves to frustrate viruses trying to latch on and perhaps to ward off attacks from our immune system. Below the sugar lies a pair of membranes, one nested in the other. The membranes block big molecules from entering *E. coli* and keep the microbe's molecules from getting out. *E. coli* depends on those molecules reacting with one another in a constant flurry. Keeping its 60 million molecules packed together lets those reactions take place quickly. Without a barrier, the molecules would wander away from one another, and *E. coli* would no longer exist.

At the same time, though, life needs a connection to the outside world. An organism must draw in new raw materials to grow, and it must flush out its poisonous waste. If it can't, it becomes a coffin. *E. coli*'s solution is to build hundreds of thousands of pores, channels, and pumps on the outer membrane. Each opening has a shape that allows only certain molecules through. Some swing open for their particular molecule, as if by password.

Once a molecule makes its way through the outer membrane, it is only half done with its journey. Between the outer and inner membranes of *E. coli* is a thin cushion of fluid, called the periplasm. The periplasm is loaded with enzymes that can disable dangerous molecules before they are able to pass through the inner membrane. They can also break down valuable molecules so that they can fit in channels embedded in the inner membrane. Meanwhile, *E. coli* can truck its waste out through other channels. Matter flows in and out of *E. coli,* but rather than making a random, lethal surge, it flows in a selective stream.

E. coli has a clever solution to one of the universal problems of life. Yet solutions have a way of creating problems of their own. *E. coli*'s barriers leave the microbe forever on the verge of exploding. Water molecules are small enough to slip in and out of its membranes. But there's not much room for water molecules inside *E. coli,* thanks to all the proteins and other big molecules. So at any moment more water molecules are trying to get into the microbe than are trying to get out. The force of this incoming water creates an enormous pressure inside *E. coli,* several times higher than the pressure of the atmosphere. Even a small hole is big enough to make *E. coli* explode. If you prick us, we bleed, but if you prick *E. coli,* it blasts.

One way *E. coli* defends against its self-imposed pressure is with a corset. It creates an interlocking set of molecules that form a mesh that floats between the inner and outer membranes. The corset (known as the peptidoglycan layer) has the strength to withstand the force of the incoming water. *E. coli* also dispatches a small army of enzymes to the membranes to repair any molecules damaged by acid, radiation, or other abuse. In order to grow, it must continually rebuild its membranes and peptidoglycan layer, carefully inserting new molecules without ever leaving a gap for even a moment.

E. coli's quandary is one we face as well. Our own cells carefully regulate the flow of matter through their walls. Our bodies use skin as a barrier, which must also be pierced with holes—for sweat glands, ear canals, and so on. Damaged old skin cells slough off as the underlying ones grow and divide. So do the cells of the lining of our digestive tract, which is essentially just an interior skin. This quick turnover allows our barriers to heal quickly and fend off infection. But it also creates its own danger. Each time a cell divides, it runs a small risk of mutating and turning cancerous. It's not surprising, then, that skin cancer and colon cancer are among the most common forms of the disease. Humans and *E. coli* alike must pay a price to avoid becoming a blur.

THE RIVER THAT RUNS UPHILL

Barriers and genes are essential to life, but life cannot survive with barriers and genes alone. Put DNA in a membrane, and you create nothing more than a dead bubble. Life also needs a way to draw in molecules and energy, to transform them into more of itself. It needs a metabolism.

Metabolisms are made up of hundreds of chemical reactions. Each reaction may be relatively simple: an enzyme may do nothing more than pull a hydrogen atom off a molecule, for instance. But that molecule is then ready to be grabbed by another enzyme that will rework it in another way, and so on through a chain of reactions that can become hideously intricate—merging with other chains, branching in two, or looping back in a circle. The first species whose metabolism scientists mapped in fine detail was *E. coli.*

It took them the better part of the twentieth century. To uncover its pathways, they manipulated it in many ways, such as feeding it radioactive food so that they could trace atoms as *E. coli* passed them from molecule to molecule. It was slow, tough, unglamorous work. After James Watson and Francis Crick discovered the structure of DNA, their photograph appeared in *Life* magazine: two scientists flanking a tall, bare sculpture. There was no picture of the scientists who collectively mapped *E. coli*'s metabolism. It would have been a bad photograph anyway: hundreds of people packed around a diagram crisscrossed with so many arrows that it looked vaguely like a cat's hairball. But for those who know how to read that diagram, *E. coli*'s metabolism has a hidden elegance.

The chemical reactions that make up *E. coli*'s metabolism don't happen spontaneously, just as an egg does not boil itself. It takes energy to join atoms together, as well as to break them apart. *E. coli* gets its energy in two ways. One is by turning its membranes into a battery. The other is by capturing the energy in its food.

Among the channels that decorate *E. coli*'s membranes are pumps that hurl positively charged protons out of the microbe. *E. coli* gives itself a negative charge in the process, attracting positively charged atoms that happen to be in its neighborhood. It draws some of them into special channels that can capture energy from their movement, like an electric version of a waterwheel. *E. coli* stores that energy in the atomic bonds of a molecule called adenosine triphosphate, or ATP.

ATP molecules float through *E. coli* like portable energy packs. When *E. coli*'s enzymes need extra energy to drive a reaction, they grab ATP and draw out the energy stored in the bonds between its atoms. *E. coli* uses the energy it gets from its membrane battery to get more energy from its food. With the help of ATP, its enzymes can break down sugar, cutting its bonds and storing the energy in still more ATP. It does not unleash all the energy in a sugar molecule at once. If it did, most of that energy would be lost in heat. Rather than burning up a bonfire of sugar, *E. coli* makes surgical nicks, step by step, in order to release manageable bursts of energy.

E. coli uses some of this energy to build new molecules. Along with the sugar it breaks down, it also needs a few minerals. But it has to work hard to get even the trace amounts it requires. *E. coli* needs iron to live, for example, but iron is exquisitely scarce. In a living host most iron is tucked away inside cells. What little there is outside the cells is usually bound up in other molecules, which will not surrender it easily. *E. coli* has to fight for iron by building iron-stealing molecules, called siderophores, and pumping them out into its surroundings. As the siderophores drift along, they sometimes bump into iron-bearing molecules. When they do, they pry away the iron atom and then slide back into *E. coli*. Once inside, the siderophores unfold to release their treasure.

While iron is essential to *E. coli*, it's also a poison. Once inside the microbe, a free iron atom can seize oxygen atoms from water molecules, turning them into hydrogen peroxide, which in turn will attack *E. coli*'s DNA. *E. coli* defends itself with proteins that scoop up iron as soon as it arrives and store it away in deep pockets. A single one of these proteins can safely hold 5,000 iron atoms, which it carefully dispenses, one atom at a time, as the microbe needs them.

Iron is not the only danger *E. coli*'s metabolism poses to itself. Even the proteins it builds can become poisonous. Acid, radiation, and other sorts of damage can deform proteins, causing them to stop working as they should. The mangled proteins wreak havoc, jamming the smooth assembly line of chemistry *E. coli* depends on for survival. They can even attack other proteins. *E. coli* protects itself from itself by building a team of assassins—proteins whose sole function is to destroy old proteins. Once an old protein has been minced into amino acids, it becomes a supply of raw ingredients for new proteins. Life and death, food and poison—all teeter together on a delicate fulcrum inside *E. coli*.

As *E. coli* juggles iron, captures energy, and transforms sugar into complex molecules, it seems to defy the universe. There's a powerful drive throughout the universe, known as entropy, that pushes order toward disorder. Elegant snowflakes melt into drops of water. Teacups shatter. *E. coli* seems to push against the universe, assembling atoms into intricate proteins and genes and preserving that orderliness from one generation to the next. It's like a river that flows uphill.

E. coli is not really so defiant. It is not sealed off from the rest of the universe. It does indeed reduce its own entropy, but only by consuming energy it gets from outside. And while *E. coli* increases its own internal order, it adds to the entropy of the universe with its heat and waste. On balance, *E. coli* actually increases entropy, but it manages to bob on the rising tide.

E. coli's metabolism is something of a microcosm of life as a whole. Most living things ultimately get their energy from the sun. Plants and photosynthetic microbes capture light and use its energy to grow. Other species eat the photosynthesizers, and still other species eat them in turn. *E. coli* sits relatively high up in this food web, feeding on the sugars made by mammals and birds. It gets eaten in turn, its molecules transformed into predatory bacteria or viruses, which get eaten as well. This flow of energy gives rise to forests and other ecosystems, all of which unload their entropy on the rest of the universe. Sunlight strikes the planet, heat rises from it, and a planet full of life—an *E. coli* for the Earth—sustains itself on the flow.

A SENSE OF WHERE YOU ARE

Life's list grows longer. It stores information in genes. It needs barriers to stay alive. It captures energy and food to build new living matter. But if life cannot find that food, it will not survive for long. Living things need to move—to fly, squirm, drift, send tendrils up gutter spouts. And to make sure they're going in the right direction, most living things have to decide where to go.

We humans use 100 billion neurons bundled in our heads to make that decision. Our senses funnel rivers of information to the brain, and it responds with signals that control the movements of our bodies. *E. coli,* on the other hand, has no brain. It has no nervous system. It is, in fact, thousands of times smaller than a single human nerve cell. And yet it is not oblivious to its world. It can harvest information and manufacture decisions, such as where it should go next.

E. coli swims like a spastic submarine. Along the sides of its cigar-shaped body it sprouts about half a dozen propellers. They're shaped like whips, trailing far behind the microbe. Each tail (or, as microbiologists call it, flagellum) has a flexible hook at its base, which is anchored to a motor. The motor, a wheel-shaped cluster of proteins, can spin 250 times a second, powered by protons that flow through its pores into the microbe's interior.

Most of the time, *E. coli*'s motors turn counterclockwise, and when they do their flagella all bundle together into a cable. They behave so neatly because each flagellum is slightly twisted in the same direction, like the ribbons on a barber's pole. The cable of flagella spin together, pushing against the surrounding fluid in the process, driving the microbe forward.

E. coli can swim ten times its body length in a second. The fastest human swimmers can move only two body lengths in that time. And *E. coli* wins this race with a handicap, because the physics of water is different for microbes than for large animals like us. For *E. coli,* water is as viscous as mineral oil. When it stops swimming, it comes to a halt in a millionth of a second. *E. coli* does not stop on a dime. It stops on an atom.

About every second or so, *E. coli* throws its motors in reverse and hurls itself into a tumble. When its motors spin clockwise, the flagella can no longer slide comfortably over one another. Now their twists cause them to push apart; their neat braid flies out in all directions. It now looks more like a fright wig than a barber's pole. The tumble lasts only a tenth of a second as *E. coli* turns its motors counterclockwise once more. The flagella fold together again, and the microbe swims off.

The first scientist to get a good look at how *E. coli* swims was Howard Berg, a Harvard biophysicist. In the early 1970s, Berg built a microscope that could follow a single *E. coli* as it traveled around a drop of water. Each tumble left *E. coli* pointing in a new random direction.

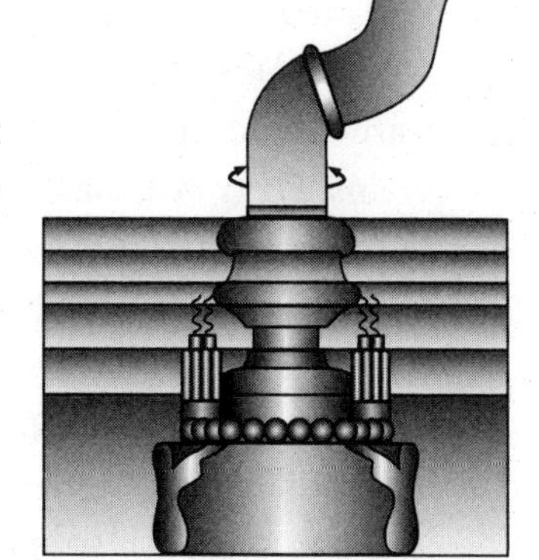

E. coli's flagellum is driven by motorlike proteins that spin in its membrane.

Berg drew a single microbe's path over the course of a few minutes and ended up with a tangle, like a ball of yarn in zero gravity. For all its busy swimming, Berg found, *E. coli* manages to wander only within a tiny space, getting nowhere fast.

Offer *E. coli* a taste of something interesting, however, and it will give chase. *E. coli*'s ability to navigate is remarkable when you consider how little it has to work with. It cannot wheel and bank a pair of wings. All it can do is swim in a straight line or tumble. And it can get very little information about its surroundings. It cannot consult an atlas. It can only sense the molecules it happens to bump into in its wanderings. But *E. coli* makes good use of what little it has. With a few elegant rules, it gets where it needs to go.

E. coli builds sensors and inserts them in its membranes so that their outer ends reach up like periscopes. Several thousand sensors cluster together at the microbe's front tip, where they act like a microbial tongue. They come in five types, each able to grab certain kinds of molecules. Some types attract *E. coli,* and some repel it. An attractive molecule, such as the amino acid serine, sets in motion a series of chemical reactions inside the microbe with a simple result: *E. coli* swims longer between its tumbles. It will keep swimming in longer runs as long as it senses that the concentration of serine is rising. If its tumbles send it away from the source of serine, its swims become shorter. This bias is enough to direct *E. coli* slowly but reliably toward the serine. Once it gets to the source, it stays there by switching back to its aimless wandering.

Scientists began piecing together *E. coli*'s system of sensing and swimming in the 1960s. They chose *E. coli*'s system because they thought it would be easy. They could take advantage of the long tradition of using mutant *E. coli* to study how proteins work. And once they had solved *E. coli*'s information processors, they would be able to take what they had learned and apply it to more complex processors, including our own brains. Forty years later they understand *E. coli*'s signaling system more thoroughly than that of any other species. Some parts of *E. coli*'s system turned out to be simple after all. *E. coli* does not have to compute barrel rolls or spiral dives. Its swim-and-tumble strategy works very well. Every *E. coli* may not get exactly where it needs to go, but many of them will. They will be able to survive and reproduce and pass the run-and-tumble strategy on to their offspring. That is all the success a microbe needs.

Yet in some important ways, *E. coli*'s navigation defies understanding. Its microbial tongue can detect astonishingly tiny changes in the concentration of molecules it cares about, down to one part in a thousand. The microbe is able to amplify these faint signals in a way that scientists have not yet discovered. It's possible that *E. coli*'s receptors are working together. As one receptor twists, it causes neighboring receptors to twist as well. *E. coli* may even be able to integrate different kinds of information at the same time—oxygen climbing, nickel falling, glucose wafting by. Its array of receptors may turn out to be far more than just a microbial tongue. It may be more like a brain.

THE MYTH OF THE TANGLED SPAGHETTI

E. coli's brainy tongue does not fit well into the traditional picture of bacteria as primitive, simple creatures. Well into the twentieth century, bacteria remained saddled with a reputation as relics of life's earliest stages. They were supposedly nothing more than bags of enzymes with some loose DNA tossed in like a bowl of tangled spaghetti. "Higher" organisms, on the other hand—including animals, plants, fungi—were seen as having marvelously organized cells. They all keep their DNA neatly wound up around spool-shaped proteins and bundled together into chromosomes. The chromosomes are tucked into a nucleus. The cells have other compartments, in which they carry out other jobs, such as generating energy or putting the finishing touches on proteins. The cells themselves have structure, thanks to a skeletal network of fibers crisscrossing their girth.

The contrast between these two kinds of cells—sloppy and neat—seemed so stark in the mid-1900s that scientists used it to divide all of life into two great groups. All species that carried a nucleus were eukaryotes, meaning "true kernels" in Greek. All other species—including *E. coli*—were now prokaryotes. Before the kernel there were prokaryotes, primitive and disorganized. Only later did eukaryotes evolve, bringing order to the world.

There's a kernel of truth to this story. The last common ancestor of all living things almost certainly didn't have a nucleus. It probably looked vaguely like today's prokaryotes. Eukaryotes split off from prokaryotes more than 3 billion years ago, and only later did they acquire a full-fledged nucleus and other distinctive features. But it is all too easy to see more differences between prokaryotes and eukaryotes than actually exist. The organization of eukaryotes jumps out at the eye. It is easy to see the chromosomes in a human cell, the intricately folded Golgi apparatus, the sausage-shaped mitochondria. The geography is obvious. But prokaryotes, it turns out, have a geography as well. They keep their molecules carefully organized, but scientists have only recently begun to discover the keys to that order.

Many of those keys were first discovered in *E. coli*. *E. coli* must grapple with several organizational nightmares in order to survive, but none so big as keeping its DNA in order. Its chromosome is a thousand times longer than the microbe itself. If it were packed carelessly into the microbe's interior, its double helix structure would coil in on itself like twisted string, creating an awful snarl. It would be impossible for the microbe's gene-reading enzymes to make head or tail of such a molecule.

There's another reason why *E. coli* must take special care of its DNA: the molecule is exquisitely vulnerable to attack. As the microbe turns food into energy, its waste includes charged atoms, which can crash into DNA, creating nicks in the strands. Water molecules are attracted to nicks, where they rip the bonds between the two DNA strands, pulling the chromosome apart like a zipper.

Only in the past few years have scientists begun to see how *E. coli* organizes its DNA. Their experiments suggest that it folds its chromosome into hundreds of loops, held in place by tweezerlike proteins. Each loop twists in on itself, but the tweezers prevent the coiling from spreading to the rest of the chromosome. When *E. coli* needs to read a particular gene, a cluster of proteins moves to the loop where the gene resides. It pulls the two strands of DNA apart, allowing other proteins to slide along one of the strands and produce an RNA copy of the gene. Still other proteins keep the strands apart so that they won't snarl and tangle during the copying. Once the RNA molecule has been built, the proteins close the strands of the DNA again. *E. coli*'s tweezers also make the damage from unzipping DNA easier to manage. When a nick appears in the DNA, only a single loop will come undone because the tweezers keep the damage from spreading farther. *E. coli* can then use repair enzymes to stitch up the wounded loop.

E. coli faces a far bigger challenge to its order when it reproduces. To reproduce, it must create a copy of its DNA, pull those chromosomes to either end of its interior, and slice itself in half. Yet *E. coli* can do all of that with almost perfect accuracy in as little as twenty minutes.

The first step in building a new *E. coli*—copying more than a million base pairs of DNA—begins when two dozen different kinds of enzymes swoop down on a single spot along *E. coli*'s chromosome. Some of them pull the two strands of DNA apart while others grip the strands to prevent them from twisting away or collapsing back on each other. Two squadrons of enzymes begin marching down each strand, grabbing loose molecules to build it a partner. The squadrons can add a thousand new bases to a DNA strand every second. They manage this speed despite running into heavy traffic along the way. Sometimes they encounter the sticky tweezers that keep DNA in order; scientists suspect that the tweezers must open to let the replication squadrons pass through, then close again. The squadrons also end up stuck behind other proteins that are slowly copying genes into RNA and must wait patiently until they finish up and fall away before racing off again. Despite these obstacles, the DNA-building squadrons are not just fast but awesomely accurate. In every 10 billion bases they add, they may leave just a single error behind.

As these enzymes race around *E. coli*'s DNA, two new chromosomes form and move to either end of the microbe. Although scientists have learned a great deal about how *E. coli* copies its DNA, they still debate how exactly the chromosomes move. Perhaps they are pulled, perhaps they are pushed. However they move, they remain tethered like two links in a chain. A special enzyme handles the final step of snipping them apart and sealing each back together. Once liberated, the chromosomes finish moving apart, and *E. coli* can begin to divide itself in two.

The microbe must slice itself precisely, in both space and time. If it starts dividing before its chromosomes have moved away, it will cut them into pieces. If it splits itself too far toward either end, one of its offspring will have a pair of chromosomes and the other will have none. These disasters almost never take place. *E. coli* nearly always divides itself almost precisely at its midpoint, and almost always after its two chromosomes are safely tucked away at either end.

A few types of proteins work together to create this precise dance. When *E. coli* is ready to divide, a protein called FtsZ begins to form a ring along the interior wall of the microbe at midcell. It attracts other proteins, which then begin to close the ring. Some proteins act like winches, helping to drag the chromosomes away from the closing ring. Others add extra membrane molecules to seal the ends of the two new microbes.

FtsZ proteins form their ring without consulting a map of the microbe, without measuring it with a ruler. Instead, it appears that FtsZ is forced by other proteins to form the ring at midcell. Another protein, called MinD, forms into spirals that grow along the inside wall of the microbe. The MinD spiral can scrape off any FtsZ it encounters attached to the wall. But the MinD spiral itself is fleeting. Another protein attaches to the back end of the spiral and pulls the MinD proteins off the wall one at a time.

A pattern emerges: the MinD spiral grows from one end toward the middle but falls apart before it gets there. The dislodged MinD proteins float around the cell and begin to form a new spiral at the other end. But as the MinD spiral grows toward the middle again, its back end gets destroyed once more. The MinD spiral bounces back and forth, taking about a minute to move from one end of the microbe to the other.

The bouncing MinD spiral scrapes away FtsZ from most of the cell. Only in the middle can FtsZ have any hope of forming the ring. And even there FtsZ is blocked most of the time by the chromosome and its attendant proteins. Only after the chromosome has been duplicated and the two copies are moving away from the middle is there enough room for FtsZ to take hold and start cutting the microbe in two.

E. coli may not have the obvious anatomy of a eukaryote cell, but it has a structure nevertheless. It is a geography of rhythms, a map of flux.

OFF THE CLIFF

E. coli caught Theodor Escherich's eye thanks to its gift for multiplication—the way a single microbe can give rise to a massive, luxurious growth in a matter of hours. If the bacteria Escherich discovered had continued to reproduce at that rapid rate, they would have soon filled his flasks with a solid microbial mass. In a few days they could have taken over the Earth. But *E. coli* did something else. It began to grow more slowly, and then, within a day, it stopped.

All living things could, in theory, take over the planet. But we do not wade through forests of puffballs or oceans of fleas. A species' exponential growth quickly slams into the harsh reality of this finite world. As *E. coli*'s population grows denser, the bacteria use up oxygen faster than fresh supplies can arrive. Their waste builds up around them, turning toxic. This collision with reality can be fatal. As *E. coli* runs out of its essential nutrients, its ribosomes get sloppy, producing misshapen protein that attacks other molecules. The catastrophe can ripple out across the entire microbe. To continue to grow under such stress would be suicidal, like driving a car over a cliff.

Instead, *E. coli* slams on the brakes. In a matter of seconds it stops reading its genes and destroys all the proteins it's in the midst of building. It enters a zombielike state called the stationary phase. The microbe begins to make proteins to defend against heat, acid, and other insults even as it stops making the enzymes necessary for feeding. To keep dangerous molecules from slipping through its membranes, *E. coli* closes off many of its pores. To protect its DNA, *E. coli* folds it into a kind of crystalline sandwich. All of these preparations demand a lot of energy, which the microbe can no longer get from food. So *E. coli* eats itself, dismantling some of its own energy-rich molecules. It even cannibalizes many of its ribosomes, so it can no longer make new proteins.

The threats faced by a starving *E. coli* are much like the ones our own cells face as we get old. Aging human cells suffer the same sorts of damage to their genes and ribosomes. People who suffer Alzheimer's disease develop tangles of misshapen proteins in their brains—proteins that are deformed in much the same way some proteins in starving *E. coli* are deformed. Life not only grows and reproduces. It also decays.

Although humans and microbes face the same ravages of time, it's the microbe that comes out the winner. If scientists pluck out a single *E. coli* in a stationary phase and put it in a flask of fresh broth, it will unpack its DNA, build new proteins, and resume its life with stately grace. Scientists can leave a colony of *E. coli* in a stationary phase for five years and still resurrect some viable microbes. We humans never get such a second chance.

Three

THE SYSTEM

TURNING ON THE GENE

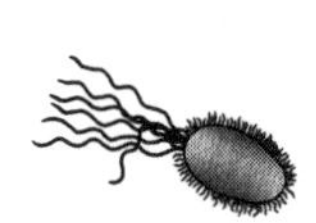

ONE DAY IN JULY 1958, François Jacob squirmed in a Paris movie theater. His wife, Lise, could tell that an idea was struggling to come out. The two of them walked out of the theater and headed for home.

"I think I've just thought up something important," François said to Lise.

"Tell!" she said.

Her husband believed, as he later wrote, that he had reached "the very essence of things." He had gotten a glimpse of how genes work together to make life possible.

Jacob had been hoping for a moment like this for a long time. Originally trained as a surgeon, he had fled Paris when the Nazis swept across France. For the next four years he served in a medical company in the Allied campaigns, mostly in North Africa. Wounds from a bomb blast ended his plans of becoming a surgeon, and after the war he wandered Paris unsure of what to do with his life. Working in an antibiotics lab, Jacob became enchanted with scientific research. But he did not simply want to find a new drug. Jacob decided he would try to understand "the core of life." In 1950, he joined a team of biologists at the Pasteur Institute who were toiling away on *E. coli* and other bacteria in the institute's attic.

Jacob did not have a particular plan for his research when he ascended into the attic, but he ended up studying two examples of one major biological puzzle: why genes sometimes make proteins and sometimes don't. For several years, Jacob investigated prophages, the viruses that disappear into their *E. coli* host, only to reappear generations later. Working with Élie Wollman, Jacob demonstrated that prophages actually insert their genes into *E. coli*'s own DNA. They allowed prophage-infected bacteria to mate with uninfected ones and then spun them apart. If the microbes stopped mating too soon, they could not transfer the prophage. The experiments revealed that the prophage consistently inserts itself in one spot in *E. coli*'s chromosome. The virus's genes are nestled in among those of its host, and yet they remain silent for generations.

E. coli offered Jacob another opportunity to study genes that sometimes make proteins and sometimes don't. To eat a particular kind of sugar, *E. coli* needs to make the right enzymes. In order to eat lactose, the sugar in milk, *E. coli* needs an enzyme called beta-galactosidase, which can cut lactose into pieces. Jacob's colleague at the Pasteur Institute, Jacques Monod, found that if he fed *E. coli* glucose—a much better source of energy for *E. coli* than lactose—it made only a tiny amount of beta-galactosidase. If he added lactose to the bacteria, it still didn't make much of the enzyme. Only after the bacteria had eaten all the glucose did it start to produce beta-galactosidase in earnest.

No one at the time had a good explanation for how genes in *E. coli* or its prophages could be quiet one moment and busy the next. Many scientists had assumed that cells simply churned out a steady supply of all their proteins all the time. To explain *E. coli*'s reaction to lactose, they suggested that the microbe actually made a steady stream of beta-galactosidase. Only when *E. coli* came into contact with lactose did the enzymes change their shape so that they could begin to break the sugar down.

Jacob, Monod, and their colleagues at the Pasteur Institute began a series of experiments to figure out the truth. They isolated mutant *E. coli* that failed to eat lactose in interesting ways. One mutant could not digest lactose, despite having a normal gene for beta-galactosidase. The scientists realized that *E. coli* used more than one gene to eat lactose. One of those genes encoded a channel in the microbe's membranes that could suck in the sugar.

Strangest of all the mutants Jacob and Monod discovered were ones that produced beta-galactosidase and permease all the time, regardless of whether there was any lactose to digest. The scientists reasoned that *E. coli* carries some other molecule that normally prevents the genes for beta-galactosidase and permease from becoming active. It became known as the repressor. But Jacob and his colleagues had not been able to say how the repressor keeps genes quiet.

In the darkness of the Paris movie theater, Jacob hit on an answer. The repressor is a protein that clamps on to *E. coli*'s DNA, blocking the production of proteins from the genes for beta-galactosidase and the other genes involved in feeding on lactose. A signal, like a switch on a circuit, causes the repressor to stop shutting down the genes.

Another similar repressor might keep the genes of prophages silent as well, Jacob thought. Perhaps these circuits are common in all living things. "I no longer feel mediocre or even mortal," he wrote.

But when François tried to sketch out his ideas for his wife, he was disappointed.

"You've already told me that," Lise said. "It's been known for a long time, hasn't it?"

Jacob's idea was so elegantly simple that it seemed obvious to anyone other than a biologist. Yet it represented a new way of thinking about life. Genes do not work in isolation. They work in circuits. Over the next few weeks, Jacob tried to explain his idea to his fellow biologists, without arousing much interest. It was not until Monod returned to Paris in the fall that Jacob found a receptive audience. The two of them began to draw circuit diagrams on a blackboard, with arrows running from inputs to outputs.

In the fall of 1958, Monod and Jacob launched a new series of experiments to test Jacob's circuit hypothesis. The experiments produced the results Jacob expected, but it would take years of research by other scientists to work out many of the details. The lactose-digesting genes are lined up next to each other on *E. coli*'s chromosome. The repressor protein clamps down on a stretch of DNA at the front end of the genes, where it blocks the path of gene-reading enzymes. With the repressor in place, *E. coli* cannot feed on lactose.

The best way to get the repressor away from the lactose-digesting genes is to give *E. coli* some lactose. Once inside the microbe, the sugar changes shape so that it can grab the repressor. It drags the repressor off *E. coli*'s DNA, allowing the gene-reading enzymes to make their way through the lactose-digesting genes. *E. coli* can then make the enzymes it needs to feed on lactose.

But *E. coli* needs a second signal to ramp up its production of beta-galactosidase: it needs to know that its supply of glucose has run out. The signal is a protein called CRP, which builds up inside *E. coli* when the microbe begins to starve. CRP grabs on to another stretch of DNA, next to the lactose-digesting genes. It bends the DNA to attract the gene-reading enzymes. Once CRP clamps on, *E. coli* begins producing lactose-digesting enzymes at top speed. If the repressor is an off switch, CRP is an on switch.

Jacob and his colleagues christened the lactose-digesting genes the *lac* operon, *operon* meaning a set of genes that are all regulated by the same switches. As Jacob suspected, operons represent a common theme in the way genes work. Hundreds of *E. coli*'s genes are arrayed in operons, each controlled by switches. Some operons carry several switches, all of which must be thrown for them to make proteins. A single protein may be able to trigger a cascade of genes, switching on genes for making more switches, allowing *E. coli* to make hundreds of new kinds of proteins.

On-off switches are everywhere in nature. Prophages remain dormant inside *E. coli* thanks to repressors that keep their genes shut down. Stress causes the repressors to fall off and the prophages to make new viruses. Operons can be found in other bacteria as well. In animals like ourselves, operons appear to be much less common. But even genes that do not sit next to each other on our genome can be switched on by the same master-control protein.

It is only through the switching on and off of genes that our cells can behave differently from one another, despite carrying an identical genome. They can form liver cells or spit out bone, catch light or feel heat. By learning how *E. coli* drinks milk, Jacob and his colleagues opened the way to understanding why we humans are more than just amoebas.

LIVING CIRCUITS

To an engineer, a circuit is an arrangement of wires, resistors, and other parts, all laid out to produce an output from an input. Circuits in a Geiger counter create a crackle when they detect radioactivity. A room is cast in darkness when a light switch is turned off. Genes operate according to a similar logic. A genetic circuit has its own inputs and outputs. The *lac* operon works only if it receives two inputs: a signal that *E. coli* has run out of glucose and another signal that there's lactose to eat. Its output is the proteins *E. coli* needs to break down the lactose.

E. coli has no wires that scientists can pull apart to learn how its circuits work. Instead, they must do experiments of the sort Jacob and Monod carried out. They observe how quickly the bacteria respond to their environment, how quickly they make a certain protein or clear another one away. Scientists combine the results of experiment after experiment into models, which they use to make predictions about how future experiments will turn out. The fundamental discoveries that Jacob, Monod, and others made about *E. coli* have led other scientists to pick apart the circuitry of other species, including us. But in the fifty years since Jacob squirmed in a cinema seat, scientists have continued to pay close attention to *E. coli.* They discovered intriguing patterns in *E. coli*'s circuitry, which they mapped out in more detail than in of any other species, and they've discovered that *E. coli*'s circuitry mimics the sort of circuitry you might find in digital cameras or satellite radios.

To prove that I'm not dabbling in idle metaphor, I want to probe the wiring of one of *E. coli*'s many circuits. This particular circuit controls the construction of *E. coli*'s flagella. It has taken the work of many scientists over many years to discover most of the genes that belong to this circuit. But in 2005, Uri Alon and his colleagues at the Weizmann Institute of Science in Rehovot, Israel, figured out what the circuit does. It acts as a noise filter.

Engineers use noise filters to block static in phone lines, blurring in images, and any other input that obscures a true signal. In the case of *E. coli,* the noise is made up of misleading cues about its environment. With the help of a noise filter it can pay attention only to the cues that matter. It's important for *E. coli* to ignore noise when it builds a flagellum because the process is a lot like building a cathedral.

The microbe must switch on about fifty genes, which make tens of thousands of proteins. Those proteins must come together in a tightly choreographed assembly. First the motor must insert itself in the membranes. A syringe has to slide through the center of the motor, which then injects thousands of proteins into the growing tail. The proteins squirm through the hollow shaft and emerge to form its new tip. The process takes an hour or two, which for *E. coli* can mean several generations. A new microbe inherits a partially built tail and passes it on, still unfinished, to its descendants.

By the time *E. coli* has finished building these flagella, the crisis may be long over. All that energy will have gone to waste. So *E. coli* keeps tabs on its surroundings, and if life does seem to be getting better, it stops building its flagella. The only problem with this strategy is that a sign of better times may actually be a fleeting mirage. If *E. coli* abandons its flagella when a single oxygen molecule drifts by, it may end up stranded in a very dangerous place. To *E. coli* these false signs are noise it must filter out of its circuits.

To explain how *E. coli* filters out noise, I will draw a wiring diagram. An arrow with a plus sign means that a signal or a gene boosts the activity of another gene. A minus sign means that the supply of protein is reduced. The first link in this circuit is from the outside world to the inner world of *E. coli.* When the microbe senses danger, it sometimes responds by producing a protein called FlhDC.

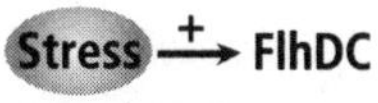

Stress —+→ **FlhDC**

FlhDC is one of *E. coli*'s master switches. It can latch on to many spots along *E. coli*'s chromosome, where it can switch on a number of genes. These genes make many of the proteins that combine to make flagella.

Stress —+→ **FlhDC** —+→ **Flagella genes** —+→ **Flagella**

In this simple form, *E. coli*'s flagella-building circuit has a major flaw. It can turn on flagella-building genes in response to stress, but it also has to shut them down as soon as the stress goes away. Once the microbe stops making new FlhDC, the old copies of FlhDC gradually disappear. As they do, the genes FlhDC controls can no longer make their proteins. The complex assembly of flagella comes screeching to a halt in response to the slightest improvement. When conditions turn bad again, this circuit has to fire up its flagella machine from scratch. In a crisis, those delays could be fatal.

E. coli does not fall victim to false alarms, however, because it has extra loops in its genetic circuit. In addition to switching on flagella genes, FlhDC switches on a backup gene called FliA.

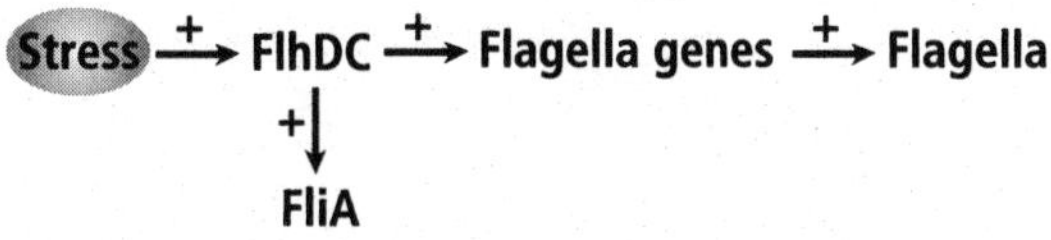

Stress —+→ **FlhDC** —+→ **Flagella genes** —+→ **Flagella**
+↓
FliA

FliA can switch on the flagella genes as well.

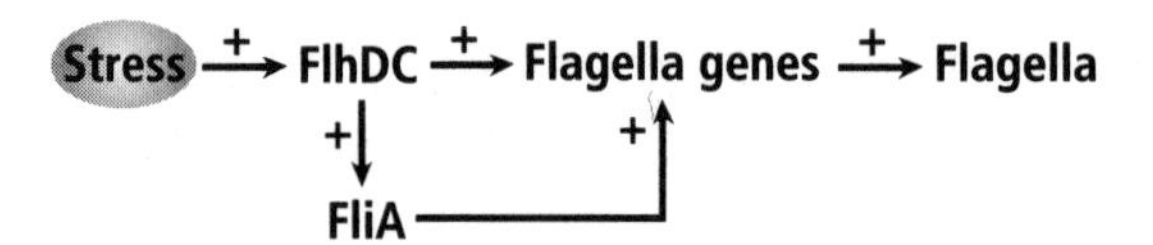

But FliA is also controlled by another protein, called FlgM. It grabs new copies of FliA as soon as *E. coli* makes them, preventing them from switching on the flagella genes. Here is the circuit with FlgM added:

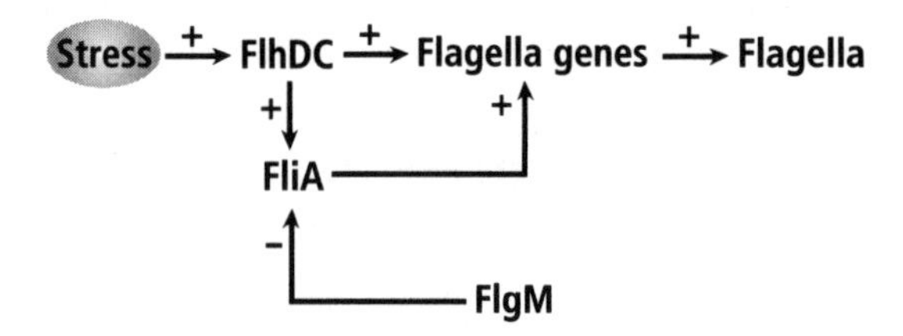

FlgM cannot keep FliA repressed for long, however, because *E. coli* can expel it through the same syringe it uses to build its flagella. As the number of FlgM proteins dwindles, more FliA genes become free to switch on the flagella-building genes.

Here, at last, is the full noise filter as reconstructed by Alon and his colleagues:

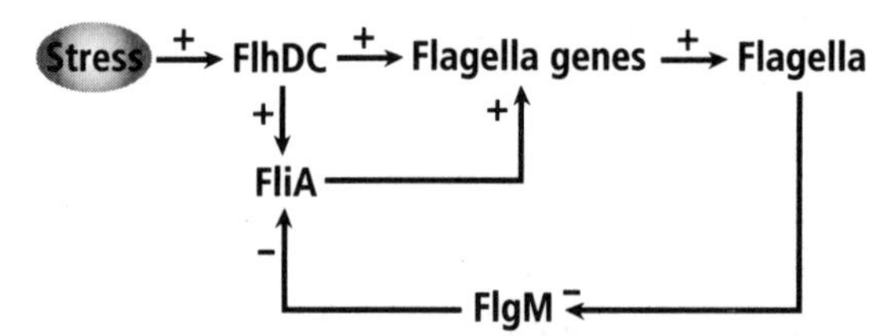

This elegant network gives *E. coli* the best of all worlds. When it starts building flagella, it remains very sensitive to any sign that stress is going away. That's because FlhDC alone is keeping the flagella-building genes switched on. But once *E. coli* has built a syringe and begins to pump out FlgM, the noise filters kick in. If the stress drops, so does the level of FlhDC. But *E. coli* has created enough free FliA genes to keep its flagella-building genes switched on for more than an hour. If the respite is temporary, *E. coli* will start making new copies of FlhDC, and its construction of flagella will go on smoothly.

E. coli can filter out noise, but it's not deaf. If conditions get significantly better, *E. coli* can stop making flagella. Its extra supply of FliA cannot last forever. The proteins become damaged and are destroyed by *E. coli*'s molecular garbage crews. If the stress does not return in time, the microbe will run out of FliA, and the circuit will shut down. The good times have truly returned.

Scientists are now starting to map the circuitry of genes in other species as carefully as Alon and his colleagues have in *E. coli.* But it will take time. Scientists don't yet know enough about how the genes and proteins in those circuits build good models. In many cases, scientists know only that gene A turns on gene B and gene C, without knowing what causes it to flip the switch or what happens when it does.

But Alon has discovered a remarkable lesson even in that tiny scrap of knowledge. He and his colleagues have surveyed the genes in *E. coli* and a few other well-studied organisms—yeast, vinegar worms, flies, mice, and humans. The arrows that link them tend to form certain patterns far more often than you'd expect if they were the result of chance. *E. coli*'s noise filter, for example, belongs to a class of circuits that engineers call feed-forward loops. (The loop in the noise filter goes from FlhDC to FliA to the flagella-building genes.) Feed-forward loops are unusually common in nature, Alon and his colleagues have shown. Nature has a preference for a few other patterns as well, which also seem to allow life to take advantage of engineering tricks like the noise filter. *E. coli* and the elephant, it seems, are built not only with the same genetic code. They're also wired in much the same way.

LIFE ON AUTOPILOT

An orange winter dusk has settled in. Out my window I can see the webs of bare maple branches. Photons stream through the window and patter on the photoreceptors lining my retina. The photoreceptors produce electric signals, which they trade among themselves and then fire down the fibers of my optic nerves into the back of my brain. Signals move on through my brain, following a network made of billions of neurons linked by trillions of branches. An image emerges. I get up from my desk to turn on the lights. At first I can see nothing outside, but after a moment my eyes adjust. I can still see the trees, down to their twigs.

I must remind myself how remarkable it is that I can still see them. A moment earlier my vision was exquisitely tuned to perceiving the world at dusk. If it had stayed that way after I turned on the light, I would have been practically blinded. Fortunately my eyes and brain can retune themselves for the noonday sun or a crescent moon. If the light increases, my brain quickly tightens my irises to reduce the light coming in. When the lights go out, my pupils expand, and my retinal neurons boost the contrast between light and dark in my field of vision. An engineer would call my vision robust. In other words, it works steadily in an unsteady world.

Our bodies are robust in all sorts of ways. Our brains need a steady supply of glucose, but we don't black out if we skip dinner. Instead, our bodies unload reserves of glucose as needed. A clump of cells develops into an embryo by trading a flurry of signals to coordinate their divisions. The signals are easily disrupted, but most embryos can still turn into perfectly healthy babies. Again and again life avoids catastrophic failure and remains on course.

Until recently, scientists had no solid evidence for where life's robustness comes from. To trace robustness to its source, they needed to know living things with a deep intimacy—the same intimacy an engineer may have with an autopilot system, using its plans to carry out experiments. But the blueprints of most living things remain classified. Among the few exceptions is *E. coli.*

E. coli faces threats to its survival on a regular basis. Set a petri dish on a windowsill on a sunny day and you bring the microbes in it to the brink of disaster. In order to work properly, a protein needs to maintain its intricate origami-like folds. Overheated proteins shake themselves loose. They can no longer do the job on which *E. coli*'s survival depends.

Yet *E. coli* does not die from a few degrees of extra heat. As the temperature rises, the microbe makes molecules known as heat-shock proteins. They defend *E. coli* in two ways. Some of them embrace *E. coli*'s jittery proteins and guide them back into their proper shape. Others recognize heat-snarled proteins that have been damaged beyond repair. They slice these hopeless proteins apart, leaving harmless fragments to be recycled.

Heat-shock proteins are lifesavers, but *E. coli* can't keep a supply of them on hand for emergencies. They are among the biggest proteins in its repertoire, and to survive a blast of heat *E. coli* may need tens of thousands of them. Making heat-shock proteins in ordinary times would be like paying the local fire company to park all its trucks in your driveway just in case your house catches fire. On the other hand, when you need a fire truck, you need it fast. If *E. coli* takes too long to manufacture heat-shock proteins, it can die while it waits to be rescued.

This tricky trade-off attracted the attention of John Doyle, an engineer at the California Institute of Technology, and his colleagues. In past years, Doyle had developed a theory for designing control systems for airplanes and space shuttles. In *E. coli* he recognized a piece of natural engineering just as impressive as anything he had helped to build. He and his colleagues began to analyze its heat-shock proteins and the way *E. coli* uses them to survive.

They found that *E. coli* controls its supply of heat-shock proteins with feedback. For engineers, feedback is what happens when they allow the output of a circuit to become an input. A thermostat uses a simple form of feedback to keep the temperature of a house stable. The thermostat senses the temperature in the house and turns on the heater if it's too cold. If the temperature gets too high, it shuts the heater down.

E. coli's defense against heat works a lot like a thermostat as well. The key protein in its thermostat is called sigma 32. Even when the temperature is cool, *E. coli* is constantly reading the gene for sigma 32 and making RNA copies. But at normal temperatures the RNA folds in on itself, and so *E. coli* cannot use it to make a protein. At normal temperatures the microbe is loaded with sigma 32 RNA but no actual sigma 32 protein.

Only when *E. coli* heats up can the sigma 32 RNA uncrumple. Now the ribosomes can read it and make huge amounts of sigma 32 protein. Each sigma 32 protein quickly finds some of *E. coli*'s gene-reading enzymes and leads them to the genes for heat-shock proteins. *E. coli* thus makes tens of thousands of heat-shock proteins in a matter of minutes.

Left unchecked, however, a sudden rush of sigma 32 would be too much of a good thing. The microbe would churn out heat-shock proteins far beyond its needs. In fact, *E. coli* makes just the right number of heat-shock proteins to cope with a particular temperature. It makes more proteins for higher temperatures, fewer for cooler ones. It exerts this fine control with a series of feedback loops.

E. coli's heat-shock proteins don't just protect against heat. They also control the thermometer protein itself, sigma 32. Some of them grab sigma 32 and tuck it away in a pocket. Others cut it to pieces. In the first few moments of dangerous heat, heat-shock proteins are too busy helping unfolded proteins to attack sigma 32. But once they get the crisis under control, more and more heat-shock proteins become free to grab sigma 32. As the level of sigma 32 drops, *E. coli* makes fewer new heat-shock proteins.

This feedback helps keep *E. coli* from exploding with heat-shock proteins. It also controls the level of heat-shock proteins. If *E. coli* is merely warm rather than scorching, the heat-shock proteins quickly reduce the level of sigma 32. But as the temperature increases, they have to cope with more unfolded proteins, and thus they allow sigma 32 to remain high so that *E. coli* will produce more heat-shock proteins. And once *E. coli* cools down to a comfortable temperature, its thermostat shuts down the heat-shock proteins almost completely.

E. coli's robust self-control comes from the feedback loops built into its network. To engineers this principle is second nature. The autopilot in a Boeing 777 uses the same kinds of feedback to keep the plane level as it is buffeted by wind shears and downdrafts. In neither case does robustness come from some all-knowing consciousness. It emerges from the network itself.

THE BIG PICTURE

Put genes together into circuits and they can do much more than they could on their own. Put circuits together and you create a living thing.

In the 1940s, Edward Tatum and other scientists got the first hints of what certain genes in *E. coli* were for. As of 2007, researchers had a pretty good idea of what about 85 percent of its genes do, making *E. coli* the gold standard of genetic familiarity. Scientists have created online databases for *E. coli*'s genes, its operons, its metabolic pathways. Mysteries remain—there are forty-one enzymes drifting around inside *E. coli* for which scientists have yet to find genes, for example—but a rough portrait of *E. coli*'s entire system is emerging, the closest thing biologists have to a complete solution to any living organism.

Bernhard Palsson, a biologist at the University of California, San Diego, has overseen the construction of a model of *E. coli*'s metabolism. As of 2007, he and his colleagues had programmed a computer with information on 1,260 genes and 2,077 reactions. The computer can use this information to calculate how much carbon flows through *E. coli*'s pathways, depending on the sort of food it eats. Palsson's model does a good job of predicting how quickly *E. coli* will grow on a diet of glucose and how much carbon dioxide it will release. If Palsson switches off the oxygen, the model shunts carbon into an oxygen-free metabolic pathway, just as *E. coli* does. If Palsson leaves out a particular protein, the model metabolism rearranges itself just as the metabolism of a real mutant *E. coli* would. It predicts *E. coli*'s behavior in thousands of conditions. The model and *E. coli* alike make the best of whatever situation they face, adjusting their metabolism in order to grow as fast as they can.

How does *E. coli*'s metabolism manage to stay so supple when it is made up of hundreds of chemical reactions? With thousands of possible pathways it could choose from, why does it choose among the best few? Why doesn't the whole system simply crash? Part of the solution lies in the shape of the network itself, the very layout of its labyrinth.

When scientists map the pathways that a carbon atom can take through *E. coli*'s metabolism, the picture they see looks like a bow tie. On one side of the bow tie are the chemical reactions that draw in food and break it down. These reactions follow each other along simple pathways, a fan of incoming arrows. Eventually the arrows all converge on the bow tie's knot. There the pathways get much more complicated. The product of a reaction may get pulled into many different reactions, depending on the conditions at that moment. It is there, in the knot, that *E. coli* creates the building blocks for all its molecules. The building blocks enter the other side of the bow tie—an outgoing fan of pathways. Each pathway produces a very different sort of molecule—this one a membrane molecule, that one a piece of RNA, another one a protein. The pathways on the far side of the bow tie fan out without crossing over. A molecule on its way to becoming a protein does not become a piece of DNA.

The bow tie architecture in *E. coli* makes good engineering sense. Man-made networks, such as a telephone network or a power grid, are often laid out in a bow tie as well. A bow tie architecture lets networks run efficiently and robustly. The Internet, for example, has an incoming fan made up of signals from e-mail programs, Web browsers, and all sorts of other software, each with its own peculiar sorts of information processing. In order for this stream of data to get onto the Internet, it must first be turned into a code that obeys the Internet's protocols. These data streams move from personal computers to servers and then into a small core of routers. The signals can then flow into an outgoing fan of pathways, toward another computer, where the standard stream of data can be converted into a picture, a document, or some other peculiar form.

In both the Internet and *E. coli,* the bow tie knot allows each network to function even when parts of it fail. A mutation that destroys one metabolic reaction will not kill *E. coli* because in the knot there are other pathways onto which it can still shunt carbon. The Internet can continue sending messages even after one of the servers shuts down because it can move the messages through another pathway.

The bow tie architecture also saves energy in both systems. If *E. coli* did not have a bow tie, it would have to create a dedicated pathway of enzymes to make every molecule it needed. Each of those enzymes would require its own gene. Instead, *E. coli*'s pathways all dump their products into the same network in the knot of the bow tie. Likewise, the Internet does not have to link every computer directly to every other one, or use special codes for every kind of file it carries. In both cases this arrangement is possible only because the entire network obeys certain rules. On the Internet every message must be converted into the same data packets. In *E. coli* all energy transfers must use the same currency: ATP.

The inventors of the Internet did not realize they were creating this kind of network. They were only trying to balance cost and speed as they joined servers together. But unintentionally they created a model of *E. coli* that spans the Earth.

VIVE LA DIFFÉRENCE

We all have our own tastes. I don't understand why some people eat snails. I can't say for sure why I dislike them, but I can certainly think up a few stories. Maybe I have a certain kind of sensor on the cells of my tongue that goes into a spasm of dismay. Or maybe some network of neurons in my brain associates the taste of snails with some awful memory from my distant past. Or maybe I simply never had the opportunity to come to love snails because I grew up eating pizza and hamburgers and peanut butter. The gastronomic window has now closed.

I have no way of knowing whether any of those possibilities is true. I can't go back in time, replay my life from the moment of conception, and see if a plate of escargots served at kindergarten lunch would have made a difference. I can't clone myself a hundred times over and send my manufactured twins to foster homes in France. I am a single, useless snail-loathing datum.

My distaste for snails is a minor example of a major fact: life is full of differences. We humans differ from one another in ways too many to count. We are shy and bold, freckled and pale, truckers and hairdressers, Buddhists and Presbyterians. We get cancers in third grade and live for a century. We have fingerprints.

Scientists have only a rough understanding of how this diversity arises. We are not merely the output of software written in a programming code of DNA. As we develop in the womb, our genes interact with signals from our mothers. The environment continues to influence those genes in unpredictable ways after birth. The food we eat, the air we breathe, the traumas and joys and boredom of childhood, and all the rest have an influence on which genes become active. Our differences are not just hard to trace but a source of pride. We can produce greatness of all kinds: Babe Ruths and Frédéric Chopins, Mae Wests and Marie Curies. They are products of our complexity, of a species in which each individual carries 18,000 genes that can become 100,000 proteins, which give rise to creatures uniquely able to experience the world, to shape their lives by words, rituals, images. And this pride colors our image of *E. coli.*

Surely *E. coli* must be all nature and no nurture. A colony descended from a single ancestor is just a billion genetically identical cousins, their behavior all run through the same genetic circuits. *E. coli* is just a single cell, after all, not a body made of a trillion cells that take years to develop. *E. coli* doesn't grow up going to private school or searching for food on a garbage dump. It doesn't wonder whether it might like snails for dinner. It's just a bag of molecules. If it is genetically identical to another *E. coli,* then the two of them will live identical lives.

This may all sound plausible, but it is far from the truth. A colony of genetically identical *E. coli* is, in fact, a mob of individuals. Under identical conditions, they will behave in different ways. They have fingerprints of their own.

If you observe two genetically identical *E. coli* swimming side by side, for example, one may give up while the other keeps spinning its flagella. To gauge their stamina, Daniel Koshland, a scientist at the University of California, Berkeley, glued genetically identical *E. coli* to a glass cover slip. They floated in water, tethered by their flagella. Koshland offered them a taste of aspartate, an amino acid that attracts them and motivates them to swim. Stuck to the slide, the bacteria could only pirouette. Koshland found that some of the clones twirled twice as long as others.

E. coli expresses its individuality in other ways. In a colony of genetically identical clones, some will produce sticky hairs on their surface, and some will not. In a rapidly breeding colony, a few individual microbes will stop growing, entering a peculiar state of suspended animation. In a colony of *E. coli,* some clones like milk sugar, and others don't.

These differing tastes for lactose first came to light in 1957. Aaron Novick and Milton Weiner, two biologists at the University of Chicago, looked at how individual *E. coli,* respond to the presence of lactose. They fed *E. coli* a lactoselike molecule that could also trigger the bacteria to make beta-galactosidase. At low levels only a tiny fraction of the microbes responded by producing beta-galactosidase. Most did nothing.

Novick and Weiner added more of the lactose mimic. The eager individuals remained eager. The reluctant ones remained reluctant. Only after the lactose mimic rose above a threshold did the reluctant microbes change. Suddenly they produced beta-galactosidase as quickly as the eager microbes.

Somehow the bacteria were behaving in radically different ways even though they were all genetically identical. Novick and Weiner isolated eager and reluctant individuals and transferred them to fresh petri dishes, where they could breed new colonies of their own. Their descendants continued to behave in the same way. Eager begat eager; reluctant, reluctant. Novick and Weiner had found a legacy beyond heredity.

There's much to be learned about *E. coli* by thinking of it as a machine with circuitry that follows the fundamental rules of engineering. But only up to a point. Two Boeing 777s that are in equally good working order should behave in precisely the same way. Yet if they were like *E. coli,* one might turn south when the other turned north.

The difference between *E. coli* and the planes lies in the stuff from which they are made. Unlike wires and transistors, *E. coli*'s molecules are floppy, twitchy, and unpredictable. They work in fits and starts. In a plane, electrons stream in a steady flow through its circuits, but the molecules in *E. coli* jostle and wander. When a gene switches on, *E. coli* does not produce a smoothly increasing supply of the corresponding protein. A single *E. coli* spurts out its proteins unpredictably. If its *lac* operon turns on, it may spit out six beta-galactosidase enzymes in the first hour, or none at all.

This burstiness helps turn genetically identical *E. coli* into a crowd of individuals. Michael Elowitz, a physicist at Cal Tech, made *E. coli*'s individuality visible in an elegant experiment. He and his colleagues added an extra gene to the *lac* operon, encoding a protein that gave off light. When he triggered the bacteria to turn on the operon, they began to make the glowing proteins. But instead of glowing steadily, they flickered. Each burst of fluorescent proteins gave off a pulse of light. Some bursts were big, and some were small. And when Elowitz took a snapshot of the colony, it was not a uniform sea of light. Some microbes were dark at that moment while others shone at full strength.

These noisy bursts can produce long-term differences between genetically identical bacteria. They turn out to be responsible for making some *E. coli* eager for lactose and others reluctant. If you could peer inside a reluctant *E. coli*, you would find a repressor clamped tightly to the *lac* operon. Lactose can sometimes seep through the microbe's membrane, and it can even sometimes pry away the repressor. Once the *lac* operon is exposed, *E. coli*'s gene-reading enzymes can get to work very quickly. They make an RNA copy of the operon's genes, which is taken up by a ribosome and turned into proteins, including a beta-galactosidase enzyme.

But each *E. coli* usually contains about three repressors. They spend most of their time sliding up and down the microbe's DNA, searching for the *lac* operon. It takes only a few minutes for one of them to find it and shut down the production of beta-galactosidase. Only a tiny amount of beta-galactosidase gets made in those brief moments of liberty. And what few enzymes do get made are soon ripped apart by *E. coli*'s army of protein destroyers. Adding a little more lactose does not change the state of affairs. Too little of the sugar gets into the microbe to keep the repressors away from the *lac* operon for long. The microbe remains reluctant.

Keep increasing the lactose, however, and this reluctant microbe will suddenly turn eager. There's a threshold beyond which it produces lots of beta-galactosidase. The secret to this reversal is one of the other genes in the *lac* operon. Along with beta-galactosidase, *E. coli* makes the protein permease, which sucks lactose molecules into the microbe. When a reluctant *E. coli*'s *lac* operon switches on briefly, some of these permeases get produced. They begin pumping more lactose into the microbe, and that extra lactose can pull away more repressors. The *lac* operon can turn on for longer periods before a repressor can shut it down again, and so it makes more proteins—both beta-galactosidase for digesting lactose and permease for pumping in more lactose. A positive feedback sets in: more permease leads to more lactose, which leads to more permease, which leads to more lactose. The feedback drives *E. coli* into a new state, in which it produces beta-galactosidase and digests lactose as fast as it can.

Once it becomes eager, *E. coli* will resist changing back. If the concentration of lactose drops, the microbe will still pump in lactose at a high rate, thanks to all the permease channels it has built. It can supply itself with enough lactose to keep the repressors away from the operon so that it can continue making beta-galactosidase and permease. Only if the lactose concentrations drop below a critical level do the repressors suddenly get the upper hand. Then they shut the operon down, and the microbe turns off.

This sticky switch helps to make sense of Novick and Weiner's strange experiments. Two genetically identical *E. coli* can respond differently to the same level of lactose because they have different histories. The reluctant one resists being switched on while the eager one resists being switched off. And both kinds can pass on their state to their offspring. They don't bequeath different genes to their descendants. Some give their offspring a lot of permeases on their membranes and a lot of lactose molecules floating through their interiors. Others give their offspring neither.

Combine this peculiar switch with *E. coli*'s unpredictable bursts and you have a recipe for individuality. If a colony of *E. coli* encounters some lactose, some of the bacteria will respond with a huge burst of proteins from their *lac* operon. They will push themselves over the threshold from reluctant to eager, and they will stay that way even if the lactose drops. Other *E. coli* will respond to the lactose with no proteins at all. They will remain reluctant. These clones take on different personalities thanks to chance alone.

E. coli also gets some of its personality from an extra layer of heredity. Some of its DNA is covered with caps made of hydrogen and carbon atoms. These caps, known as methyl groups, change the response of

E. coli's genes to incoming signals. They can, in effect, shut a gene down for a microbe's entire life without harming the gene itself. When *E. coli* divides in two, it bequeaths its pattern of methyl groups to its offspring. But under certain conditions, *E. coli* will pull methyl groups off its DNA and put new groups on—for reasons scientists don't yet understand.

Some of the factors that spin the wheel for *E. coli* spin it for us as well. To smell, for example, we depend on hundreds of different receptors on the nerve endings in our noses. Each neuron makes only one type of receptor. Which receptor it makes seems to be a matter of chance, determined by the unpredictable bursts of proteins within each neuron. Our DNA carries methyl groups as well, and over our lifetime their pattern can change. Pure chance may be responsible for some changes; nutrients and toxins may trigger others. Identical twins may have identical genes, but their methyl groups are distinctive by the time they are born and become increasingly different as the years pass. As the patterns change, people become more or less vulnerable to cancer or other diseases. This experience may be the reason why identical twins often die many years apart. They are not identical after all.

These different patterns are also one reason why clones of humans and animals can never be perfect replicas. In 2002, scientists in Texas reported that they had used DNA from a calico cat named Rainbow to create the first cloned kitten, which they named Cc. But Cc is not a carbon copy of Rainbow. Rainbow is white with splotches of brown, tan, and gold. Cc has gray stripes. Rainbow is shy. Cc is outgoing. Rainbow is heavy, and Cc is sleek. New methylation patterns probably account for some of those differences. Clones may also get hit by a unique series of protein bursts. The very molecules that make them up turn them into individuals in their own right.

At the very least, *E. coli*'s individuality should be a warning to those who would put human nature down to any sort of simple genetic determinism. Living things are more than just programs run by genetic software. Even in minuscule microbes, the same genes and the same genetic network can lead to different fates.

Four

THE *E. COLI* WATCHER'S FIELD GUIDE

A HUMAN KRAKATAU

ON AUGUST 26, 1883, a little world was born. An island volcano called Krakatau, located between Java and Sumatra in the Sunda Strait, hurled a column of ash twenty miles into the air. Rock turned to vapor and roared across the strait at 300 miles an hour. The eruption left a submerged pit where the cone of the volcano had been, along with a few lifeless islands. Nine months later, a naturalist who visited the scene reported that the only living thing he could find was a single small spider.

The new islands of Krakatau lay twenty-seven miles from the nearest land. It took years for life to make its way across the water and take hold again. A film of blue-green algae grew over the ash. Ferns and mosses sprouted. By the 1890s a savanna had emerged. Along with the spiders came beetles, butterflies, and even a monitor lizard. Some of the arriving species swam to the islands, some flew, and some simply drifted on the wind.

These species did not take hold on Krakatau in a random scramble. Rugged pioneers came first and later gave way to other species. The savanna surrendered to forests. Coconut and fig trees grew. Orchids, fig wasps, and other delicate species could now move onto the islands. Early settlers such as zebra doves could no longer find a place in the food web and vanished. Even now, more than 120 years after the eruption, Krakatau is not finished with its transformation. In the future it may be ready to receive bamboo, which will revolutionize its ecosystem yet again.

The history of Krakatau followed ecological rules that guide life wherever new habitats appear. Volcanic eruptions wipe islands clean. Landslides clear mountainsides. As glaciers melt, shorelines bounce out of the sea.

And babies are born. To microbes, a newborn child is a Krakatau ready to be colonized. Its body starts out almost completely germ free, and in its first few days *E. coli* and other species of bacteria infect it. They establish a new ecosystem, which will mature and survive within the child through its entire life. And it will develop over time according to its own ecological rules.

There is much more to *E. coli*'s life than can be seen in a petri dish. Its pampered existence in the laboratory makes very few demands on it. Out of the 4,288 genes scientists have identified in *E. coli* K-12, only 303 appear to be essential for its growth in a laboratory. That does not mean the other 3,985 genes are all useless. Many help *E. coli* survive in the crowded ecosystem of the human gut, where a thousand species of microbes compete for food.

A scientist studying *E. coli* in a flask may completely overlook some of its essential strategies for surviving in the real world. For all the work that has gone into *E. coli* over the past century, for example, microbiologists often fail to acknowledge just how social a creature it is. To survive, *E. coli* work together. The bacteria communicate and cooperate. Billions of them join together to build microbial cities. They wage wars together against their enemies.

In the real world there is no single way of being an *E. coli*. *E. coli* K-12 is just one of many strains that live in warm-blooded animals and have many strategies for surviving. Some are harmless gut grazers. Others shield us from infections. And still others kill millions of people a year. To know *E. coli* by K-12 alone is a bit like knowing the family *Canidae* from a Pomeranian dozing on a silk pillow. Outside there are dingoes and bat-eared foxes, red wolves and black-backed jackals.

FINDING A HOME

E. coli is a pioneer. Long before most other microbes have moved into a human host, it has established a healthy colony. *E. coli* may infect a baby during the messy business of childbirth, hitch along on the fingertips of a doctor, or make its leap as mother nurses child. It rides waves of peristalsis into the stomach, where it must survive an acid bath. As the swarms of protons in hydrochloric acid seep into it, *E. coli* builds extra pumps that can flush most of them out. It does not try to behave like a normal microbe in the stomach; instead, it enters what one scientist has called "a Zen-like physiology." Except for the proteins it needs to defend against stomach acid, *E. coli* simply stops making proteins altogether.

After two hours in this acid Zen, *E. coli* is driven out of the stomach and into the intestines. Its pumps continue driving out its extra protons until its interior gets back its negative charge. Its biological batteries power up once more, and it can now begin to make new proteins and repair old ones. It returns to the everyday business of living. *E. coli* has not yet reached its new home, though—it must first travel through the small intestine and into the large one. The distance may be only thirty feet, but it's about 7 million times the length of *E. coli.* If you dived into the ocean in Los Angeles and swam 7 million body lengths, you could cross the Pacific.

As *E. coli* drifts through the human gut, its hook-tipped hairs snag on the intestinal walls. A gentle flow of food is enough to detach the hooks, allowing the microbe to roll along. But if the flow becomes strong, the hairs begin to grip stubbornly to the wall. It just so happens that the hairs bring *E. coli* to a halt exactly in the place in the large intestine that suits it best, where food flows by at top speed. The warmth of the gut triggers it to make proteins it can use to harvest iron, to break down sugar, and to weld together amino acids. It begins to feed and thrive, at least for a few days.

As *E. coli* grows and multiplies, it prepares the way for its own downfall. It uses up much of the oxygen in the intestines and alters their chemistry by releasing carbon dioxide and other wastes. It creates a new habitat that other species of microbes can invade and dominate. This ecosystem *E. coli* helps to build in our bodies is spectacular. It can reach a population of 100 trillion, outnumbering the cells of our body ten to one. Scientists estimate that a thousand species of microbes can coexist in a single human gut, which means that if you were to make a list of all the genes in your body, the vast majority of them would not be human.

As other species prosper, *E. coli* dwindles away until it makes up just one-tenth of 1 percent of the population of gut microbes. It becomes prey to viruses and predatory protozoans. It must compete with other microbes for food. But it also comes to depend on other species of microbes for food. As its host grows older and gives up milk, the gut starts to fill with starches and other complex sugars that *E. coli* can't break down. It's like going to a restaurant and having your waiter suddenly switch your chocolate mousse with a bowl of hay. *E. coli* must now wait for other species of bacteria to break down complex sugars so it can feed on their waste.

Yet even as a minor scavenger, *E. coli* may be able to repay the other microbes for their services. Some research suggests that by clearing away simple sugars, scavengers like *E. coli* may allow other microbes to break down complex sugars more quickly. *E. coli* also continues to snatch up what little oxygen accumulates in the gut from time to time. By keeping the level of oxygen at a steady low, *E. coli* makes the gut reliably comfortable for the vast majority of resident microbes. Cradled in this ecological web, *E. coli* colonies will grow in the human gut for the host's entire lifetime. As many as thirty different strains may live there at any moment. It is a very rare person who is ever *E. coli* free.

Here is another way in which we are like *E. coli:* we, too, depend on our microbial jungle. We need bacteria to break down many of the carbohydrates in our food. Our microbial passengers synthesize some of the vitamins and amino acids we need. They help control the calories that flow from our food to our bodies. A change in the bacteria in your gut may change your weight. Intestinal microbes also ward off diseases, a fact that has led doctors to feed premature infants protective strains of *E. coli.* The bacteria protect the gut by releasing chemicals that repel pathogens and by creating a tightly knit community that the pathogens simply can't invade.

It is difficult, in fact, to say exactly where these bacteria stop and our own immune systems begin. They help our immune systems manage a delicate balance between killing pathogens and not destroying our own tissues. Studies show that some strains of *E. coli* can cool down battle-frenzied immune cells. A healthy supply of *E. coli* may help ward off not just pathogens but autoimmune diseases such as colitis. Some scientists argue that our immune systems return the favor by stimulating the bacteria to form thick protective clusters that coat the intestines. The clusters not only block invaders but also prevent individual microbes from penetrating the lining of the gut. All this biochemical goodwill makes sense—after all, we and *E. coli* are members of the same collective.

TOGETHERNESS

In 2003, Jeffry Stock and his colleagues at Princeton University put *E. coli* in a maze. The maze, which measured less than a hundredth of an inch on each side, had walls of plastic and a roof of glass. The scientists submerged it in water and then injected *E. coli* into the entrance. The bacteria began to spin their flagella and swim. Stock's team had added a gene for a glowing protein to each *E. coli* so they could follow their trail as the microbes wandered through the labyrinth.

At first the bacteria seemed to move randomly. But they gradually gathered together and began to swim in schools. Some of the schools got trapped in a dead end, where the bacteria were content to stay with one another. The other bacteria swam after them, and after two hours the dead end was filled with a huddled mass of glowing microbes.

To figure out how the bacteria were finding one another, the Princeton scientists set mutants loose in the maze. They found that *E. coli* can congregate as long as their microbial tongues taste the amino acid serine. It just so happens that in the normal course of its metabolism, *E. coli* casts off serine in its waste. Scientists had known of the microbe's attraction to serine since the 1960s, but they had generally assumed that it had something to do with the microbe's search for food. *E. coli*'s sociable flocking in the maze raised another possibility: its tongue may be tuned to find other *E. coli.*

Not long ago *E. coli* and most other bacteria were considered loners. After all, they seemed to lack the sort of glue that holds societies together: a way to communicate. They cannot write e-mail; they cannot shake their tail feathers; they cannot sing across a desert at dawn. But *E. coli* does have a kind of language of its own and its own kind of society.

E. coli's social life has been overlooked for decades because most biologists have been more interested in the bare basics of its existence: how it feeds, grows, and reproduces. They've perfected the recipe for getting *E. coli* to do all three things as fast as possible. The warm, oxygen-rich, overfed life *E. coli* enjoys in the lab favors individual microbes that can breed quickly. But it bears little resemblance to *E. coli*'s normal existence. Although each person eats about sixty tons of food in a lifetime, *E. coli* may starve for hours or days. When it does get the chance to eat, it may be presented with a low-energy sugar barely worth the effort it takes to break down. *E. coli* may have to compete with other microbes for every molecule. At the same time, it must withstand assaults from viruses, predators, and man-made dangers such as antibiotics. Its host may become ill, devastating its entire habitat. One of the best ways to withstand all these catastrophes is to join forces with other *E. coli.*

Once they gather, the bacteria may do a number of things. Under some conditions a group of *E. coli* will sprout a new kind of flagellum, one that's far longer than its ordinary tail. The new flagella join together, tethering millions of bacteria into a single seething mass. Instead of swimming, they swarm across a surface, squirting out molecules that soak up water and create a carpet of slime. Swarming allows *E. coli* to glide across a petri dish or, scientists suspect, across an intestinal wall.

E. coli can also settle down and build a microbial city. Scientists have long been aware that bacteria can form a cloudy layer of scum on their flasks, known as a biofilm. Biofilms simply annoyed biologists at first. But a closer look revealed biofilms to be marvelously intricate structures. All microbes can make biofilms, and scientists suspect that the vast majority of microbes spend most of their lives in one. Biofilms form slimy coats on river bottoms, on the ocean floor, at the bottom of acid-drenched mine shafts, and on the inner walls of our intestines.

Biofilms may be everywhere, but studying them is not a simple matter. Scientists have had to ditch their flasks and petri dishes and think of new kinds of experiments. Some have built special chambers with warm flowing water to mimic the human gut. Under the right conditions, *E. coli* will settle down inside them and begin to build its biofilm. As the bacteria drift through the chamber, some alight on the bottom. Normally the microbes immediately let go and swim on, but sometimes they settle down instead. Some experiments suggest that *E. coli* make this decision if they detect other *E. coli* nearby. They sense their fellow microbes by the chemicals they release—not just serine and other sorts of waste but special molecules that serve as signals and can change the way other *E. coli* behave.

Once a group of *E. coli* has committed itself to forming a biofilm, the microbes start to build sticky fibers to snag one another and pull together into a tight cluster. They're joined by more floating *E. coli,* and the cluster grows. They begin to squirt a rubbery slime from their pores, entombing

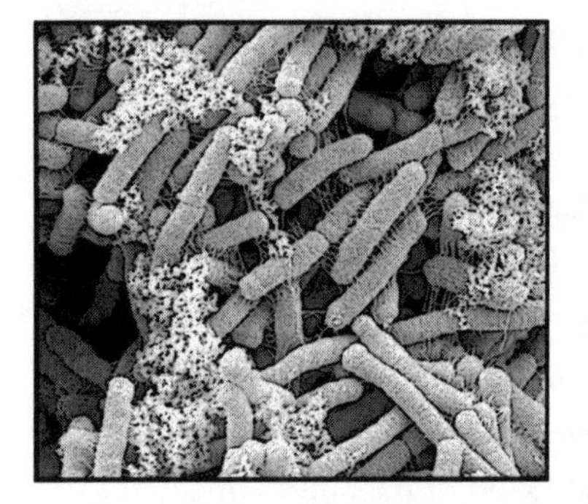

A biofilm of *E. coli*

themselves in a matrix. As the biofilm takes shape, it does not form a flat sheet. It grows looming towers, broad pedestals, and a network of crisscrossing avenues. All of these changes require each microbe to switch hundreds of genes on and off in a complicated, coordinated fashion. *E. coli* biofilms are in some ways like our bodies. A biofilm may not get up and walk around on two legs. But, like our cells, it forms collectives in which different cells take on different jobs and work together to promote their shared survival.

Scientists are still trying to figure out exactly why *E. coli* bothers to build biofilms. An individual microbe must make a great sacrifice to join the effort, spending a lot of its precious energy to build the glue that will join it to other microbes. If an individual *E. coli* should happen to get stuck deep inside the biofilm, it will have a harder time getting food than it would have if it remained floating free. These costs may be outweighed by benefits. Biofilms may provide *E. coli* with sustenance and protection. Biofilms can withstand harsh swings of the environment. Viruses may have a harder time penetrating biofilms than infecting single cells. Antibiotics are a thousand times weaker against biofilms than against individual microbes.

Biofilms may also allow bacteria to work together to catch food. Nutrients may get caught in the rubbery slime of biofilms and flow down canals to reach out-of-the-way microbes. Bacteria can also work cooperatively in biofilms by dividing their labors. The ones near the surface can get more food and oxygen than the ones buried deep inside. But they also face more stress. The *E. coli* nestled at the base of a biofilm may slip into a state of suspended animation, a kind of microbial seed bank. From time to time they may break off from the biofilm and drift away, becoming free-floating individuals or settling back down on the gut to build a new biofilm.

Humans, the supremely social species, don't cooperate just to build cities and help their fellow humans. They also cooperate to wage war. And here again *E. coli* mirrors our social life. We build missiles and bombs.

E. coli builds chemical weapons. Known as colicins, these deadly molecules kill in many ways. Some pierce the microbe's membrane like a spear, forcing its innards to spill out. Others block *E. coli* from building new proteins. Others destroy DNA.

In order to launch a colicin attack, some *E. coli* must make the ultimate sacrifice. A few microbes in a population will build hundreds of thousands of colicin molecules in a matter of seconds, until they swell with weaponry. The microbes do not have channels through which they can neatly pump out their colicins. Instead, they make suicide enzymes that cut open their membranes. As they explode, the colicins blast out and hit neighboring *E. coli.* Their close relatives are spared the attack, however, because they carry genes that produce a colicin-disabling antidote. The sacrifice of a few *E. coli* clears away the competition, and their fellow clones prosper. The social life of *E. coli,* it seems, extends beyond life itself.

FAREWELL, MY HOST

Once a strain of *E. coli* establishes itself in our guts, it can remain there for decades. But the bacteria also escape their hosts, by a route that's so obvious there's no need to dwell on it. Suffice it to say that every day the world's human population releases more than a billion trillion *E. coli* into the environment. Countless more escape from other mammals and from birds. They are swept down sewer pipes and streams, sowed upon the ground and sea. They must withstand summers and winters, droughts and floods. They must eke out an existence without a luxurious diet of half-digested sugar. For long stretches they may have to survive with no food at all. In soil and water there are many predators waiting to devour *E. coli,* including nematode worms and creeping amoebas. Some predators overpower *E. coli* by sheer size. Others, such as the bacteria *Bdellovibrio,* push their way into *E. coli*'s periplasm and destroy it from within. The bacteria *Myxococcus xanthus* release molecules that smell to *E. coli* like the whiff of food. The unlucky microbe swims to its own destruction.

Leaving their hosts is probably a quick trip to death for most *E. coli.* But life can handle bad odds. Oaks shower the ground with acorns, almost none of which survive to become saplings. Our own bodies are made of trillions of cells, only a few of which may escape our own death by giving rise to children. Even if only a tiny fraction of *E. coli* in the wild survives and manages to find a new host, its life cycle will continue. And *E. coli* has several tricks for surviving on the outside. Its versatile metabolism lets it feed on many carbon-bearing molecules—even TNT. If a soil predator tries to eat it, the microbe can avoid being digested and instead thrive as a parasite. And if worse comes to worst, *E. coli* can fold down its DNA into a rugged crystal, slip into the stationary phase, and survive for years.

Or, just perhaps, *E. coli* can abandon hosts altogether. From time to time, scientists discover populations of *E. coli* that appear to be thriving as full-time outdoor microbes. In Australia, for example, researchers have discovered huge blooms of *E. coli* in lakes where none had been expected. The lakes are free of fecal matter, receiving no sewage or farm runoff. Yet on a warm day they are loaded with millions of billions of *E. coli.* The bacteria seem different from more familiar strains. For one thing they make an unusually tough capsule, which may act as a microbial wet suit, allowing them to survive year-round in the lakes. They no longer need hosts to avoid extinction. They have broken free.

A DAY AT THE FAIR

In central Connecticut, where I live, agricultural fairs are serious business. Every summer one town after another—Goshen, Durham, Haddam Neck—raises tents and Ferris wheels. Trailers arrive, rattling along the rocky paths, full of oxen ready to drag concrete blocks. Mayors and selectmen are summoned for cow-milking contests. The fairs have survived long after the agricultural communities that produced them wilted away. Yet they still swarm with thousands of people who come to see prize goats, delicately wrought pies, and flouncing roosters.

I go with my wife and two daughters to a few fairs each summer, and each time we go, I lose my sense of time. I feel as if I'm back in an age when a typical ten-year-old would know how to shear a sheep. But just when I've almost completely lost my moorings in a tent full of livestock, I notice a wooden post staked in the ground by the entrance, holding a box of soap. It snaps me back to the twenty-first century, and when we leave the tent I make very sure my daughters scrub their hands.

These tents are home to some exquisitely vicious bacteria. The microbes live in the animals winning the ribbons at the fairs, and they fall with the droppings into the hay, float off on motes of dust, hitchhike on the bristles of flies. They spread through the tents, sticking to floors and fences and wool and feathers. It takes a tiny dose of them—just a dozen entering the mouth—to make a person hideously ill. The intestines bleed; kidneys fail. Antibiotics only make the attack worse. All doctors can do is hook their patients to an intravenous line of saline solution and hope for the best. Most people do eventually recover, but some will suffer for the rest of their lives. A few will die.

When pathologists test the fatal bacteria, they meet up with a familiar friend: *E. coli.*

E. coli comes in many strains. All of them share the same underlying biology, but they range enormously in how they make a living. Most are harmless, but outside laboratories, *E. coli* also comes in forms that can sicken or kill. To know *E. coli,* to know what it means for it to be alive, it's not enough to study a tame strain such as K-12. The deadly strains are members of the species as well.

Scientists did not appreciate how dangerous *E. coli* could be for decades after Theodor Escherich discovered the bacteria. The first clear evidence that not all strains of *E. coli* were harmless bystanders came in 1945. John Bray, a British pathologist, had been searching for the cause of "summer diarrhea," a deadly childhood disease that swept across Britain and many other industrialized countries every year. Bray hunted for bacteria that were common in sick children and missing from healthy ones.

Bray searched for the bacteria with antibodies, the best tools of his day. Antibodies are made by our immune cells when they encounter proteins from a pathogen. The antibodies can then attack the pathogen by recognizing its protein. Because antibodies are so exquisitely specific to their targets, they will ignore just about any other protein they encounter. Bray created antibodies to pathogens such as *Salmonella* by injecting the bacteria into a rabbit. Once the rabbit's immune system had mounted an attack, Bray extracted the antibodies from its blood. He then added the antibodies to cultures of bacteria he reared from the diarrhea of sick children. He wanted to see if they would reveal any pathogens. They did not.

As Bray puzzled over what kind of antibodies to make next, a pediatrician mentioned to him that children sick with summer diarrhea give off a semen-like smell. Bray knew that was also the smell of certain strains of

E. coli. So he made antibodies to *E. coli* and added them to his cultures. They immediately found their targets. Bray found that 95 percent of the sick children responded to his antibody test. Only 4 percent of the healthy children did.

Bray had identified only a single strain of disease-causing *E. coli,* but in later years scientists would identify many others. Some had long been known to medicine, but under different names. In 1897, Kiyoshi Shiga, a Japanese bacteriologist, discovered the cause of a form of bloody diarrhea called bacillary dysentery. It had *E. coli*'s basic rod-shaped anatomy, but Shiga did not call it *E. coli.* After all, many other species were rod shaped as well. And Shiga's microbe produced a cell-killing toxin that no one had ever observed *E. coli* make. In addition, *E. coli* could digest lactose, the sugar in milk, but Shiga's bacteria could not. These differences and others like them caused Shiga to declare it a species of its own, which later scientists named in his honor: *Shigella.* Only in the 1990s, when scientists could examine *Shigella*'s genes letter by letter, did they realize that it was just a strain—actually, several strains—of *E. coli.*

As the years passed, scientists discovered still more strains of *E. coli* that could cause diseases. Some strains attacked the large intestine. Others attacked the small intestine. Some lived harmlessly in the gut but could cause painful infections if they got into the bladder, sometimes creeping all the way up to the kidneys. Other strains cause lethal blood infections, and still others reach the brain and cause meningitis. The scale of their cruelty is hard to fathom. *Shigella* alone strikes 165 million people every year, killing 1.1 million of them. Most of the dead are children. I can only wonder what Theodor Escherich would have thought if he had discovered that many of the bacteria killing his young patients were actually his *Bacterium coli communis.*

Although many strains of *E. coli* are deadly, one has earned more headlines in recent years than all the rest combined. It goes by the name of O157:H7, a code for the molecules on its surface. *E. coli* O157:H7 is the strain that makes petting zoos hot zones, that can turn spinach or hamburger into poison, that can cause organ failure and death. For all its notoriety, though, it's relatively new to science.

In February and March 1982, 25 people in Medford, Oregon, developed cramps and bloody diarrhea. Doctors identified a strain of *E. coli* in some of the patients that had never been seen before. Three months later the same strain caused an outbreak in Traverse City, Michigan. The source of the bacteria proved to be undercooked hamburgers that the victims had eaten at McDonald's restaurants. A pattern had emerged, and now scientists began to hunt for *E. coli* O157:H7 in samples of bacteria taken from patients in earlier years. Out of 3,000 *E. coli* strains collected from American patients in previous years, 1 proved to be O157:H7. It came from a woman in California in 1975. Searches in Great Britain and Canada turned up 7 more cases, but none before 1975.

O157:H7 slipped back into obscurity for a decade. It emerged again in the mid-1990s in a series of outbreaks across the world. In 1993, an outbreak spread in undercooked restaurant hamburgers in Washington State sickened 732 people. Four of them died. Scientists found that cows, sheep, and other livestock can carry O157:H7 in their intestines without getting sick. An estimated 28 percent of cows in the United States carry O157:H7. It can move from animal to human through bad butchering. If a cow's colon is nicked during slaughter, the bacteria can get mixed into the meat. As meat from many cows gets blended together, *E. coli* O157:H7 can spread through tons of beef. Most of the bacteria are killed off by cooking. But a single crumb of raw beef can carry enough *E. coli* O157:H7 to start a dangerous infection.

Vegetarians are not safe either. Cows shed *E. coli* O157:H7 in their manure, and once on the ground the microbe can survive for months. On farms the bacteria can spread from manure to crops, possibly carried by slugs and earthworms or ferried by irrigation. In 1997, radish sprouts tainted with O157:H7 sickened 12,000 people in Japan, killing 3. Today in the United States the vegetable-growing business is almost as industrialized as the beef business, with a few massive companies supplying produce across much of the country. They are also extending the reach of *E. coli* O157:H7. In September 2006, contaminated spinach from a single farm made people sick across the country, striking 205 people in twenty-six states. Three months later, it was lettuce distributed to Taco Bell restaurants in five states, striking 71 people.

When *E. coli* O157:H7 first passes the lips of one of its human victims, it does not seem much different from a harmless strain. Only after it has drifted through the stomach and reached the large intestine does it begin to show its true colors. *E. coli* O157:H7 has an unusual ability to eavesdrop on us. The cells of the human intestines produce hormones, and the microbe has receptors that can grab them. The hormones tell the bacteria that it's time to prepare to make us sick. They build themselves flagella and swim, scanning the molecules floating by for signals released by their fellow *E. coli* O157:H7. They follow the signals and gather together. Once they've formed a large enough army, they begin constructing their weapons.

Their most potent weapon is a syringe they use to pierce intestinal cells and inject a cocktail of molecules. The molecules reprogram the cells. The skeleton-like fibers that give the cells structure begin sliding over one another. A pedestal-like cup rises from the top of each cell, giving *E. coli* O157:H7 a place to rest. The cells begin to leak, and the bacteria feed on the passing debris. Along with diarrhea comes bleeding, and *E. coli* O157:H7 snatches up the iron in the blood with siderophores.

It's at about this point, about three days after ingesting *E. coli* O157:H7, that people start to feel awful. They develop violent diarrhea, which begins to turn bloody. The cramps can feel like knife stabs. Most people infected with *E. coli* O157:H7 can recover within a few days. But for every twenty people who get infected, one or two have much worse in store. Their *E. coli* O157:H7 releases a new kind of toxin. This one invades cells and attacks their ribosomes, the factories that build proteins. The cells die and burst open. The toxins move from the intestines into the surrounding blood vessels and spread to the rest of the body. They trigger blood clots and seizures. They shut down entire organs, particularly the kidneys. For some the toxin is fatal. Even for the lucky ones, recovery can take years. Some will need dialysis for the rest of their lives. Children may suffer brain damage and have to learn how to read again.

E. coli O157:H7 gets a lot of press because it can create sudden epidemics in industrialized countries, but it is just one of many dangerous strains that can make us sick in many ways. *Shigella,* for example, does not rest on a pedestal the way *E. coli* O157:H7 does. It wanders. Once it reaches the intestines, it releases molecules that loosen the junctions between the cells that make up the gut wall and slips through one of the gaps. The breach draws the attention of nearby immune cells, which crawl after the microbe. But *Shigella* does nothing to camouflage itself. On the contrary, it goes out of its way to produce molecules that provoke a strong attack.

The immune cells chase after *Shigella* and devour it. But instead of killing *Shigella,* the immune cells are killed by their prey. *Shigella* releases molecules that trigger the immune cells to commit suicide and burst open. The dying immune cells draw the attention of living ones, but they are equally helpless to stop *Shigella.* In fact, they only make it easier for more *Shigella* to invade, by opening up more gaps in the intestinal wall as they push their way in.

Having fended off the immune system, *Shigella* chooses a cell in the intestinal wall to invade. It builds itself a syringe very much like the one made by *E. coli* O157:H7 and pierces a cell. The molecules it injects do not cause the cell to form a pedestal but, rather, cause it to open a passageway through which *Shigella* can slither. Once inside, it takes control of the cell's skeleton. It moves forward by causing one of the cell's fibers to grow from its back end while it hacks apart the fibers that cross its path. Once *Shigella* has finished feasting on the cell, it pushes its way out through the membrane and invades a neighbor. The dying cell summons more immune cells to the infection, and they open up more gaps through which more *Shigella* invade.

How is it that *E. coli* can be so many different things? We tend to assume that a species is made up of individuals that all share the same essence. In the ways *Shigella* and *E. coli* O157:H7 act, they seem like completely different species from the harmless K-12. Yet a comparison of their DNA shows otherwise.

If you should find yourself scrubbing your hands outside a livestock tent, stop for a minute and look around. Consider the chickens on display, showing off their chandeliers of feathers. Observe the rabbits burdened with ears too big to lift, the enormous pigs obediently following humans on leashes. Think of their less ridiculous cousins: the jungle fowl, the jackrabbit, the wild boar. These animals demonstrate that there are no fixed essences in life. One of the most important rules of life is that it changes. Boars become pigs, and harmless *E. coli* become killers. It just so happens that *E. coli* is one of the best guides to how life evolves—over days, decades, and billions of years. It vindicates Charles Darwin's central insights, yet it also reveals how much more bizarre and more fascinating evolution can be than Darwin ever anticipated.

Five

EVERFLUX

THAWING OUT THE ANCESTORS

IN A CORNER OF A LABORATORY at Michigan State University, a table rocks in a precise circle. On top of the orbital shaker are a dozen flasks filled with broth. The liquid swirls without ever breaking a ripple. Each flask contains millions of *E. coli.* They are tended by a biologist named Richard Lenski and his team of technicians and students. Lenski's experiment looks like countless other experiments that are taking place around the world, but there is one important difference. A typical experiment with *E. coli* may last only a few hours. A team of scientists might use that time to run the bacteria through a maze or rear them without oxygen to see which genes they switch on and off. Once the scientists get enough data to see a pattern, they write up the results and dump the bacteria. But the experiment in Richard Lenski's lab began in 1988, and forty thousand generations later it's still going.

Lenski launched the experiment with a single *E. coli.* He placed it on a sterile petri dish and let it divide into identical clones. These clones then became the founders of twelve separate—but genetically identical—lines. Lenski put each line into its own small flask. Instead of the endless feast of sugar that *E. coli* normally enjoys in laboratories, Lenski put his microbes on starvation rations. The bacteria ran out of their glucose by the afternoon. The following morning, Lenski transferred 1 percent of the surviving *E. coli* to a new flask with a fresh supply of sugar.

Periodically Lenski and his students drew some bacteria from each flask and stored them in a freezer. The bacteria's descendants went on multiplying daily. From time to time, Lenski has thawed out some of the early ancestors. He allows them to recover from their freeze, start eating again, and begin reproducing. And then he has compared them with their descendants. In the process, Lenski has discovered something significant: the bacteria are not what they once were. They are twice as big as their ancestors. They reproduce 70 percent faster. They've also become picky about the food they eat. If they're fed any sugar other than glucose, they grow more slowly than their ancestors. And some of them now mutate at a far higher rate than before. The descendants, in other words, have evolved into something measurably different from their ancestors.

In *The Origin of Species,* Charles Darwin wrote that "natural selection will always act with extreme slowness, I fully admit." With *E. coli,* Lenski has done something Darwin never dared dream of: he has observed evolution in his own time.

LAMARCK ON THE BEACH

I live close to the Long Island Sound, and from time to time my wife and I take our girls down to the water. The girls throw rocks and gather seaweed. On some days we are joined by nervous sandpipers. They skitter across the beach, stopping to jab their beaks into the mud before skittering off again on their pencil legs.

Two centuries ago, on a beach on the other side of the Atlantic, a French naturalist watched wading birds as well, and he wondered how they had come to be. Jean-Baptiste Lamarck concluded that they had gradually changed over generations to adapt to their environment. They had evolved. In 1801, he described the evolution of wading birds this way:

> One may perceive that the bird of the shore, which does not at all like to swim, and which however, needs to draw near to the water to find its prey, will be continually exposed to sinking in the mud. Wishing to avoid immersing its body in the liquid, it acquires the habit of stretching and elongating its legs. The result of this for the generations of these birds that continue to live in this manner is that the individuals will find themselves elevated as on stilts, on long naked legs.

"Wishing" is only a crude translation of what Lamarck had in mind. He pictured a "subtle fluid" coursing through birds and all other living things, animating them and controlling their growth and movements.

This subtle fluid was influenced by the habits the animals acquired as they explored the world. As a giraffe stretched for a leaf on a tree, the subtle fluid coursed into its neck. As more and more fluid traveled through it, the neck grew longer. Likewise, a wading bird stretched its legs to extract itself from the mud. It grew longer legs. Giraffes and wading birds alike could pass their altered bodies to their offspring.

Lamarck did not believe he was terribly original on this point. "The law of nature by which new individuals receive all that has been acquired in organization during the lifetime of their parents is so true, so striking, so much attested by the facts, that there is no observer who has been unable to convince himself of its reality," he wrote.

And yet as common as that perception may have been, today Lamarck alone is linked to it. That's because he described this change more provocatively than anyone else before him, making it part of an ambitious theory to explain the evolution of all of life's diversity. Life, Lamarck argued, was forced to change by an inherent drive from simplicity to complexity. That drive has turned microbes into animals and plants. And at each stage of the rise of complexity, species have also acquired traits they need for their particular environment and have passed them down to their offspring.

Lamarck died in 1829, poor, blind, and scorned for his theory. But he raised questions that naturalists could not shake off: how to explain the fossil record, for example, and the distribution of similar species around the world. Thirty years after Lamarck's death, Charles Darwin offered his own explanation. He argued for evolution, but he dismissed Lamarck's inexorable drive from simplicity to complexity. Darwin instead argued that life evolved primarily by natural selection.

Each generation of a species contains a vast range of variations. In the case of shorebirds, some individuals have long legs and some have short ones. Some of those variations allow individuals to survive and reproduce more successfully than others. They pass down their traits to their offspring, and generation after generation their traits become more and more common. Over millions of years, natural selection can produce a wide range of bodies. In birds, for example, feet might evolve into the striking talons of eagles, the webbed flippers of ducks, and the slender poles that keep sandpipers from sinking into the mud. Natural selection acts only on the legs the birds are born with, not on any changes the birds might experience during their life.

By the late 1800s most biologists recognized the reality of evolution, but they were divided as to how life evolves. Many accepted natural selection, but others preferred something along the lines of Lamarck. The German biologist August Weismann wanted Lamarck banished from biology. He made his case by rearing mice and cutting off their tails along the way. Over many generations, the mice never grew shorter tails as a result. Neo-Lamarckians dismissed Weismann's experiments as meaningless. The animals had not needed shorter tails, they argued, so they never produced them. The neo-Lamarckians doubted the power of natural selection, claiming that the fossil record revealed long-term trends in the history of life that short-term natural selection could not produce.

The followers of Darwin and Lamarck clashed for decades. Uncertainty kept the fights going, because scientists could not get a close look at the chemistry behind heredity. They needed an organism they could observe reproducing and acquiring an adaptation generation by generation. What they needed, it turned out, was *E. coli.*

SLOT MACHINES AND VELVET STAMPS

One night in 1942 in Bloomington, Indiana, an Italian refugee sat in a country club, teasing a friend at a slot machine.

The refugee was named Salvador Luria. He had trained as a doctor in Turin, but when he discovered viruses and bacteria he abandoned his medical career for research. During World War II he fled Italy for Paris, where he joined the scientists at the Pasteur Institute studying *E. coli* and its viruses. As the Germans closed in on Paris, Luria fled again, this time to New York. In the United States he met his hero, Max Delbrück, and the two began to work together. The scientists explored the life cycles of viruses as the viruses slipped in and out of *E. coli.* They collaborated with scientists working with the newly invented electron microscope to spy on the creatures as they invaded their hosts. And for several years, Luria and Delbrück puzzled over how *E. coli* recovers from the plagues visited on it by scientists.

In a typical experiment, researchers would add viruses to a dish full of bacteria, and the bacteria would completely disappear from view. But the viruses did not kill them all. After a few hours the survivors would produce visible colonies once more. The bacteria in the new colonies were all resistant; if the scientists moved them into fresh petri dishes and exposed them to the same viruses, their offspring would resist infection, too.

This sort of behavior in bacteria turned a lot of microbiologists into neo-Lamarckians. *E. coli* seemed to respond to viruses the same way shorebirds responded to mud. The challenge had caused them to acquire resistance, which they could then pass on to their descendants. Other experiments seemed to fit this pattern as well. When scientists switched *E. coli*'s diet from glucose to lactose, it began to produce the enzyme necessary for feeding on lactose, as did its descendants. And one other factor also made many microbiologists into neo-Lamarckians: there was little evidence that bacteria had genes. As far as many microbiologists could tell, a microbe such as *E. coli* was nothing but a bag of enzymes and other molecules that could react to changes in its environment.

But some microbiologists thought otherwise. They argued that bacteria did have genes, and that, like the genes of animals, these could mutate spontaneously. In some cases, a mutation might, through pure luck, give a microbe an advantage, such as resistance to a virus. According to this rival explanation, *E. coli* followed Darwin's rules, not Lamarck's.

No one had put the alternatives to a good test, and Luria and Delbrück spent months puzzling over how they might do so. They had failed to come up with an experiment by 1942, when they parted ways after Luria accepted a job at Indiana University, "a place I had never heard of," he wrote later. Not long afterward Luria found himself in Bloomington sitting next to a colleague who was playing a slot machine. The professor was losing, and when Luria teased him he stalked off.

"Right then I began giving some thought to the actual numerology of slot machines," Luria wrote in his autobiography.

The slot machine the professor was playing was programmed to deliver only a few big jackpots. It might have been built differently. It might have provided the same small chance of paying out a jackpot on every pull of the arm. In that case the jackpot would have given out many more prizes, but much smaller ones. Suddenly Luria realized he had figured out how to run an experiment on *E. coli*'s resistance that could test Darwin's theory versus Lamarck's.

The next day Luria began rearing flasks of *E. coli*. Each flask started out with just a few hundred microbes. Since resistant *E. coli* are extremely rare—about one in a million—the founders of each flask were all almost certainly vulnerable. Any resistance to viruses would appear in the flask only after its population began to grow.

After the *E. coli* populations had grown for a while, Luria took some bacteria from each one and spread them on petri dishes laced with viruses. He waited for epidemics to strike, and then for resistant *E. coli* colonies to emerge.

According to Lamarck, living things acquire new traits as they face new challenges, then pass these traits down to their offspring. If Luria's *E. coli* obeyed Lamarck, the bacteria would acquire resistance *after* Luria exposed them to viruses. That would mean that once Luria had inoculated his virus-laden dishes, every microbe had the same small chance of evolving resistance. Luria ought then to have discovered a few resistant colonies in every dish. The experiment would have resembled a slot machine that pays out a lot of small wins.

If *E. coli* obeyed Darwin, on the other hand, the experiment would play out like a slot machine with a few big wins. According to Darwin's followers, *E. coli* has a rare random chance of mutating every time it divides regardless of what it is experiencing. In other words, the bacteria in Luria's experiment might have acquired resistance to viruses while they were growing in the flasks, long *before* Luria exposed them to the viruses. That head start would have produced a very different result for the experiment. If a mutation had emerged early on in one of the colonies, the mutant would have had a lot of time to produce offspring. When Luria took some of the bacteria from such a colony and placed them in a petri dish with viruses, a fair number of them would already be resistant. They would grow into many new colonies in the dish.

In some of the other flasks, resistant mutants would arise much later. They would have had less time to produce offspring by the time Luria exposed them to viruses. As a result, they'd produce fewer colonies in the petri dishes. And in still other flasks, no mutants would arise at all. Their bacteria would all die, leaving their dishes empty. So instead of a few colonies growing in most dishes—the Lamarckian prediction—Darwinian mutations would produce a few dishes loaded with colonies and the rest with few or none.

Luria let his slot machines spin, and then he began to count spots.

When he was done, the verdict was clear: a few dishes were packed with colonies while many were empty. Life's slot machine had paid out a handful of big jackpots. Darwin had won.

In 1943, Luria and Delbrück published these results, which would earn them shares in a Nobel Prize in 1969. Later generations of biologists would look back at Luria's experiment as one of the greatest of the twentieth century. It provided compelling evidence that bacteria, like animals and plants, pass down their traits to their offspring through genes. It showed that those genes change spontaneously, and they can become more common in a population through natural selection. And the experiment became a powerful scientific tool: simply by counting colonies of bacteria, scientists can calculate how often mutations arise.

But when Luria and Delbrück first published the experiment, they did not bowl over the skeptics. Neo-Lamarckians remained unconvinced, pointing out that the researchers had had to rely on a lot of indirect clues to draw their conclusions. It was possible that the test tubes had not all been alike. Some might have had traces of soap or some other contamination that might have altered the bacteria. For another decade, microbiologists went on debating how bacteria adapted.

The controversy did not die until Joshua Lederberg, the scientist who discovered *E. coli* sex, tested the jackpot hypothesis with a new experiment. Lederberg and his wife, Esther, wrapped sheets of velvet around the ends of wooden cylinders that were as wide as a petri dish. The Lederbergs then stamped the velvet into a dish of *E. coli,* coating the material with the microbes, and then pressed it into a dish stocked with viruses. The Lederbergs repeated the procedure, stamping three virus-laden dishes with *E. coli* from the same original dish.

Within a few hours almost all the bacteria the Lederbergs had put in the dishes were dead from infections. A few mutants survived, however, and began to produce colonies visible to the eye. The Lederbergs photographed each dish and then looked at the pictures side by side. The constellation of mutant colonies was the same in each dish.

The Lederbergs concluded that the bacteria must have acquired mutations in the original dish. When they stamped it, the Lederbergs picked up mutants from the same spots. They transferred the bacteria to the same spot in the dishes laden with viruses. If *E. coli* had obeyed Lamarck, it would have acquired resistance only after the Lederbergs had exposed it to the viruses. There would be no reason to expect resistant bacteria to emerge in precisely the same spots in different dishes.

The Lederbergs recognized that they were seeing resistant bacteria only after they had been exposed to the viruses, so they took the experiment one step further to prove that the mutants were resistant before they encountered viruses. They pressed a velvet stamp into a dish that contained just a few colonies and then pressed it into a dish full of viruses. They waited for resistant bacteria to produce new colonies in the virus-laden dish. Each new colony corresponded to a colony in the original dish. The Lederbergs took some bacteria from the original colonies and put them in flasks, where they could breed into huge numbers. They then repeated the experiment on the new bacteria, growing a few colonies in a dish and pressing them with the velvet stamp. Now all the colonies were resistant. The Lederbergs seeded a second flask of bacteria from the dish and repeated the experiment yet again.

No matter how many times they repeated the procedure, the bacteria remained resistant to viruses even though none of them had been exposed to viruses over the course of the experiment. In 1952, the Lederbergs published their results, arguing that a few resistant bacteria had acquired mutations before the experiment began. Those bacteria had passed down the resistance gene to their descendants. To cling to Lamarck now became absurd.

These experiments on *E. coli* helped fuse evolution and genetics into a new synthesis. And as scientists continued to learn more about genes and proteins, the workings of natural selection became more clear. A mutation may change the sequence of a gene and thus the structure of its protein. In some cases, a lethal mutation might disable an essential protein. Others make no difference. And a few actually increase reproductive success. The advantage or disadvantage of a mutation depends on the environment. A mutation that confers resistance to viruses will give *E. coli* a reproductive advantage if viruses are menacing it. If not, the mutation makes no difference. It may even be a burden.

Over the past fifty years, evolutionary biologists have heaped up a mountain of evidence demonstrating that evolution does indeed take place this way. In most cases, though, they have had to study evolution indirectly, by comparing the genes of different organisms to see how natural selection has driven them apart from a common ancestor. But in a few species scientists have observed evolution as it happens, generation by generation, mutation by mutation. Among the most generous of these species is *E. coli.*

EVOLUTION UNFOLDING

When Salvador Luria ran his slot machine experiment, he captured a single round of evolution. A population of *E. coli* faced a challenge—an attack of viruses—and natural selection favored resistant mutants. But in every generation, natural selection can shape a species. New mutations arise, genes mix to form new combinations as they pass from parent to offspring, and the shifting environment creates new challenges. On this grander scale, evolution can be far harder to observe. Life has had millions of years to change, whereas scientists are on this earth for only a few decades. Darwin had resigned himself to studying evolution from a distance, and a century later most evolutionary biologists were following suit. They would compare genes in different species to learn how they diverged or search for new versions of genes that had arisen in response to new challenges. They would look for the effects of natural selection in the past. But in the 1980s a number of scientists decided to watch evolution in the present. They set out to observe *E. coli* and other bacteria undergo natural selection in their laboratories.

One of those scientists was Richard Lenski. Lenski started his scientific career hiking the Blue Ridge Mountains in search of beetles. He wanted to learn how beetles help hold together the Southern Appalachian food web. Lenski focused his work on a handful of species of *Carabus* ground beetles. He hoped to determine what controlled their population—cold snaps and heat waves perhaps, or maybe the competition for prey. The question was not just academic. The ground beetles might well be protecting the forests by keeping tree-destroying pests in check. Understanding the ecology of ground beetles might make it possible to predict outbreaks of pests and perhaps even prevent them.

Each spring, Lenski climbed the slopes and dug holes. He put plastic cups in them, covering the cups with funnels. Beetles tumbled down the funnels into the cups, and Lenski returned each day to count them. He marked the beetles and set them free. He tracked how much weight they gained each summer. He compared how many *Carabus sylvosus* he caught with how many *Carabus limbatus.* He compared how many beetles lived in dense forests with how many inhabited clear-cuts.

Lenski looked for patterns. In science, patterns become stronger the more times an experiment can be repeated. Doctors put thousands of people on an experimental drug. Physicists fire a laser millions of times to discover the ways of the photon. Ecologists also replicate their experiments when they can, but each datum demands far more labor. For his clear-cutting study, Lenski built a grand total of four enclosures, two in the clear-cut and two in the forest, each holding sixteen traps. With so few trials he could catch sight of only fleeting shadows, hazy signs of the forces governing the beetles.

Lenski came down from the mountains. He decided he would have to find another creature he could study to get some answers to the big questions on his mind. He found *E. coli.* When Lenski looked at a flask of *E. coli,* he saw a mountain. It was an ecosystem filled with billions of individual organisms. Like his beetles, *E. coli* searched for food and reproduced. They were preyed upon by viruses rather than by salamanders. *E. coli*'s ecosystem might be simpler than the Blue Ridge Mountains, but simplicity can be a virtue in science. A researcher can precisely control every variable in an experiment to see the effect of each one.

Best of all, *E. coli* is the sort of creature that can, in theory, evolve very fast. Mutations may occur only rarely, but with millions of microbes in a single flask a few mutations will arise in every generation. And because *E. coli* can reproduce in as little as twenty minutes, a beneficial mutation may let a mutant overtake a colony in a matter of days.

Lenski set up an experiment that was simple yet powerful. He gave his bacteria a limited supply of glucose and thus created a huge evolutionary pressure. Their ancestors had been fed endless meals of sugar, and they had adapted to that diet. The microbes that could convert the food to offspring fastest took over the population. In Lenski's experiment, genes that sped up breeding were no longer beneficial. His bacteria grew slowly if at all. Any new mutation that allowed the microbes to survive the conditions better, Lenski reasoned, would be strongly favored by natural selection.

As *E. coli* passed through thousands of generations in his laboratory, evolution's mark began to emerge. When Lenski pitted the ancestral bacteria against their descendants on their new diet, the new microbes reproduced faster. The more time passed, the better adapted the bacteria became. After a decade, the bacteria could grow far faster than their ancestors. The course of their evolution was not smooth: the bacteria might spend several hundred generations without any observable change, only to go through a rapid evolutionary burst. And as *E. coli* evolved to grow faster, Lenski detected other changes.

Lenski's students continue to nurture his dynasty of *E. coli* from one generation to the next, and other scientists have used similar methods to run experiments of their own. Some have watched *E. coli* adapt to life at the feverish temperature of 107 degrees Fahrenheit. Others have unleashed viruses on the bacteria and observed them become resistant, only to have the viruses evolve ways to overcome their resistance, starting the cycle all over again. While Lenski's experiment remains the longest running by far, much shorter experiments have been able to yield striking results. Bernhard Palsson and his colleagues, for example, fed five populations of *E. coli* glycerol, a carbon compound used in soaps and face creams. Ordinary *E. coli* does a lousy job of feeding on glycerol, but Palsson drove the evolution of glycerol gourmets. After only forty-four days (660 generations of *E. coli*), the bacteria could grow twice as fast as the founders of the population.

Whether it battles viruses, adapts to a diet of glycerol, or copes with heat, *E. coli* unmistakably evolves. Its swift pace of evolution in these experiments may reflect rapid evolution in the wild. After all, each time the microbe finds itself in a new environment, its evolutionary pressures suddenly shift. Genes that allow *E. coli* to thrive in a gut may mutate into forms better suited to life in the soil.

These experiments have allowed scientists to put natural selection under a microscope, teasing apart the individual mutations that benefit *E. coli.* Each time the microbe divides, it has a roughly 1-in-100,000 chance of mutating in a way that lets its descendants grow faster. The boost is often small, but it can allow a mutant's descendants to outbreed their cousins. And those mutants in turn have a small chance of picking up a second mutation that makes them even faster growers. In Palsson's 660-generation experiment, he and his colleagues confirmed two or three mutations in each population. Lenski estimated that over the course of 40,000 generations his lines have picked up as many as 100 beneficial mutations.

Beneficial mutations can take several forms. Some involve the change of a single base in a gene, something equivalent to changing *LIFT* to *LIFE.* These mutations can change the structure of a protein and thus change the way it works. It may slice a molecule more effectively than before, or start responding to a new signal. Other mutations accidentally create an extra copy of a stretch of DNA. In Palsson's experiment these duplicated segments ranged from 9 bases long to 1.3 million. Accidental duplications can create new copies of old genes. Natural selection may favor them because they produce extra proteins, which *E. coli* can use to grow and reproduce. But over time one of the copies may acquire new mutations, allowing it to take on a new function. Mutations can also snip out chunks of DNA, and microbes that lose genetic material are sometimes favored by natural selection. It's possible that proteins that were originally useful become a burden to *E. coli.*

Experiments such as these show that mutations arise randomly. And the effects of the mutations depend on how the mutations allow an organism to thrive in its own peculiar set of conditions. But does that mean evolution plays out purely by chance? The late paleontologist Stephen Jay Gould dreamed of an experiment to answer the question, which he called replaying life's tape. "You press the rewind button and, making sure you thoroughly erase everything that actually happened, go back to any time and place in the past . . . ," he wrote in his 1989 book *Wonderful Life.* "Then let the tape run again and see if the repetition looks at all like the original."

Short of time travel, Gould thought the best way a scientist could answer that question was by examining the fossil record, documenting the emergence and extinction of species. But experiments on *E. coli* can also address the question, at least on a scale of years. What makes experiments such as Lenski's particularly powerful is that evolution unfolds many times over, not just once. From an identical ancestor, Lenski produced twelve lines, each of which went through its own natural selection. Lenski and his colleagues may not be able to rewind the tape of *E. coli*'s evolution, but they can create twelve identical copies of the same tape and watch what happens when they all play at the same time.

It turns out that the tapes are not identical, nor are they entirely different. In Lenski's experiments all twelve lines grew faster than their ancestors, but some lines grew far faster than others. They all grew larger, but some became round while others remained rod-shaped. When scientists have taken a close look at the genomes of evolved bacteria, they have found many differences in their DNA. One reason evolution can take different paths is that mutations are not simple. A mutation may be beneficial in one microbe but downright harmful in another. That's because a mutated gene's effects depend in part on how it cooperates with other genes. In some cases the genes may work together well, but in other cases they may clash.

Despite those differences, natural selection can override many of the quirky details of history. While Lenski's lines may not be identical, they have tended to evolve in the same direction. They have also converged on a molecular level. Lenski and his colleagues have found several cases in which the same gene has mutated in all their lines. Even genes that have not evolved a new sequence have changed in a similar way. Some genes now make more proteins, and some make fewer. Lenski and his colleagues took a close look at how the expression of genes changed in two lines of *E. coli.* They found fifty-nine genes, and in all fifty-nine cases, the genes had changed in the same direction in both lines. The evolutionary song remains the same.

THE TANGLED BANK

"It is interesting to contemplate a tangled bank," Darwin wrote in *The Origin of Species,* "clothed with many plants of many kinds, with birds singing on the bushes, with various insects flitting about, and with worms crawling through the damp earth, and to reflect that these elaborately constructed forms, so different from each other, and dependent upon each other in so complex a manner, have all been produced by laws acting around us."

Darwin did not believe that he could see the production of life's tangled bank as it happened. Life evolves into new species over vast stretches of time, he argued, changing as slowly as mountains rise and islands sink. He could only look at the results of evolution around him, such as the distribution of related species around the world, to reconstruct the tangled bank's history. Today most scientists who study the diversity of plants and animals still follow Darwin's lead. The evidence they've amassed indicates that new species generally take thousands of years or more to branch off from other species. For the most part, it's a waste of time to sit around hoping to watch a new species emerge.

It turns out, however, that some of the same forces that drive the origin of species can be observed in a dish of *E. coli.* In the early 1990s, Julian Adams, a microbiologist at the University of Michigan, used a single microbe to found an *E. coli* population. Adams and his colleagues supplied the bacteria with a little glucose. Unlike Lenski, Adams replenished their sugar so that they never faced outright starvation. The bacteria began to evolve, adapting to the new conditions. But to Adams's surprise, natural selection did not favor a single strategy. When he put the bacteria on petri dishes, they grew in two types of colonies: some formed big splotches, and others formed small ones.

Adams thought he might have contaminated his original colony with another strain, so he shut down the experiment and started over again. After the new colony had adapted to the low-glucose diet, Adams spread the microbes on plates again. And again he discovered the same big- and small-splotch makers. Adams ran the experiment a few more times, and he found that it took about 200 generations for the two types of microbes to emerge. He realized that a single clone was evolving time and again into two distinct types of *E. coli.*

Those two types turned out to be ecological partners. The large colonies are inhabited by microbes that do a better job than their ancestors at feeding on glucose. One of the waste products they give off is acetate. *E. coli* can survive on acetate, although it grows more slowly on it than on glucose. Adams discovered that some of his *E. coli* were becoming more efficient at feeding on acetate than their ancestors were. The acetate feeders grow slowly, but they aren't driven to extinction because they are taking advantage of a food that the faster-growing bacteria aren't eating. A food chain had emerged spontaneously in Adams's lab as organisms began to depend on one another for survival.

Other scientists have confirmed Adams's findings with experiments of their own. And they've created new kinds of ecological diversity from a single *E. coli* ancestor. Instead of a glucose-only diet, Michael Doebeli and his colleagues at the University of British Columbia supplied *E. coli* with both glucose and acetate. After a thousand generations, Doebeli found that the bacteria had evolved into big and small colonies. But they were different from the big and small colonies that Adams had produced. Both colonies in Doebeli's experiment fed on glucose and acetate. The difference between them was a matter of timing. The big colonies fed on glucose until it ran out, and then they turned to acetate. The small colonies switched over sooner, so that they had a head start.

Doebeli and his colleagues then looked closely at how the genes in each colony had evolved. Typically, when *E. coli* is feeding on glucose, it keeps the genes for digesting acetate tightly repressed. If it made both sets of enzymes at the same time, they would get snared in a metabolic traffic jam. When the time comes to switch sugars, the bacteria must first destroy the enzymes for glucose and then build enzymes that can break down acetate. Doebeli found that in the small colonies, natural selection had favored mutants that stopped repressing their acetate genes. When glucose and acetate were available, these mutants fed on both kinds of sugar but did a lousy job of it compared with the glucose specialists in the large colonies. They got a reward for this sacrifice, however: they could leap quickly to take advantage of acetate while the big colony slowly retooled itself.

These experiments on *E. coli* may shed light on how new species form. Nature has formed its own petri dishes in Nicaragua, where dead volcanoes have filled with rainwater. These crater lakes are completely isolated from neighboring lakes and rivers, but on rare occasion a hurricane can sweep fish into them. In Lake Apoyo, which formed about 23,000 years ago, two species of cichlids live together. One of the fish, known as the Midas cichlid, is a big creature that roots around in the muck and crushes snails. The other fish, the arrow cichlid, is a thin, quick-darting creature that hunts for insect larvae in the open water. Their DNA indicates that the Midas cichlid was swept into the lake after it formed and that the arrow cichlid evolved from it. The split may have taken only a few thousand years.

Whether scientists study cichlids or *E. coli* or any other organism, they face the same question: Why specialize? Why don't organisms evolve to become jacks-of-all-trades instead? There may simply be limits to how well one organism can do many things. Sooner or later they encounter a trade-off. A mutation that helps *E. coli* feed on acetate may interfere with its ability to feed on glucose. By trying to do everything, generalists may lose out to specialists, which do one thing far better than anything else.

Cichlids may face similar trade-offs. A hybrid cichlid may not be particularly well adapted to eating snails or hunting for larvae, and it will have less reproductive success than the fish at the two ends of the spectrum. As more species emerge in an ecosystem, they create more opportunities for specialists to find a new way to make a living. And so over time, Darwin's bank tangles itself.

Six

DEATH AND KINDNESS

THE ANARCHIST PRINCE

CHARLES DARWIN WAS BURIED DURING a grand funeral in Westminster Abbey in 1882. Biologists were soon fighting over his legacy. In 1888, the British zoologist Thomas Huxley published a shocking essay, "Struggle for Existence and Its Bearing upon Man." In it he summoned up an ugly picture of nature as a combat of all against all. "The animal world is on about the same level as the gladiator's show," he wrote. "The creatures are fairly well-treated, and set to fight—whereby the strongest, the swiftest, and the cunningest live to fight another day. The spectator has no need to turn his thumbs down, as no quarter is given." In order to be moral, Huxley believed, humans had to work against nature.

Huxley's essay drew a stinging attack from an anarchist prince. Pyotr Alekseyevich Kropotkin was born in 1842 to a wealthy Russian nobleman. In his teenage years he served as a page to Tsar Alexander II, but he became disillusioned with the court and went to Siberia to serve in the army. There he worked as the secretary of a prison-reform committee, and the horrors he witnessed in the labor camps turned him into a radical anarchist. At the same time, he was developing into a first-rate scientist. Kropotkin joined a geographic survey in 1864 and spent the next eight years studying the Siberian landscape.

On his return from Siberia, Kropotkin soon ended up in jail for his politics. He escaped and fled to Europe, where he wrote pamphlets that earned him fame and more time in jail. Huxley's essay appeared just as Kropotkin had emerged from a three-year stint in a French prison. He settled in England, where he immediately set about writing a series of essays attacking what he saw as Huxley's distortion of both man and nature. His essays were eventually published as the best-selling book *Mutual Aid.*

Human morality is not artificial, Kropotkin argued, but in fact profoundly natural. "Sociability is as much a law of nature as mutual struggle," he wrote. Cooperation has evolved thanks to the advantages it offers over selfish behavior. Animals do not abandon one another but instead show care and concern. He recounted example after example of kindness in the animal kingdom, from horses that helped one another escape a grassland fire to horseshoe crabs righting overturned friends.

One can only wonder what Kropotkin would have thought of *E. coli*. Perhaps he would have been pleased to watch billions of microbes working together to build biofilms, to follow their swarming flocks traveling with intertwined flagella. He might have been startled by the selfless sacrifice of bacteria exploding with colicins that kill other strains. Or perhaps he would not have been startled at all. *E. coli*'s spirit of cooperation came as something of a surprise to scientists at the end of the twentieth century, but Kropotkin had written prophetic words a hundred years earlier: "Mutual aid is met with even amidst the lowest animals," he wrote, "and we must be prepared to learn some day, from the students of microscopic pond-life, facts of unconscious mutual support, even from the life of micro-organisms."

Kropotkin belonged to the same scientific era as Darwin. He was an observant nineteenth-century naturalist with no understanding of DNA and its mutations. It was not until the mid-1900s that scientists recognized how mutations arise in individuals and help them outcompete other members of their species. But when this view of evolution first emerged, many biologists recoiled from it much as Kropotkin had recoiled from Huxley's gladiatorial spectacle.

Kropotkin's intellectual grandchildren asked how competition among individuals could give rise to behavior that benefits entire groups. Fish join together into giant schools that move like a single organism. Sterile ants tend the offspring of their queen. A meerkat will stand guard so that its companions can nose around for food. If a meerkat acquired a mutation that made it stand high to keep watch over its companions, it would become easier prey. Even if natural selection could produce these selfless behaviors, biologists wondered, how could it prevent individuals from exploiting the altruism of others?

For *E. coli* the evolution of cheating is no mere thought experiment. When a colony runs out of food, the bacteria engage in a complicated cooperative dance as they enter a stationary phase. The microbes send signals to one another to synchronize their actions as they collapse their DNA and halt their production of proteins. By entering the stationary phase together, the bacteria improve the chances that at least some of them will survive until conditions improve, even though many of them may die along the way. Yet Roberto Kolter of Harvard and a former student, Marin Vulić, discovered that some bacteria do not dance to the same dying tune.

Vulić and Kolter discovered that mutants arose in their *E. coli* colony that could rouse themselves from the limbo of the stationary phase and start to feed. They fed not on sugar but on the amino acids excreted by their dormant companions. As some of the stationary bacteria died, they burst open. The mutants then fed on their proteins and DNA. The diet of the mutants was meager, but it was enough to allow them to reproduce. Over the course of several weeks the cheaters' descendants came to dominate the entire population.

This betrayal was not a rare fluke. Time and again when Vulić and Kolter starved *E. coli,* cheaters evolved and thrived. They did so according to the fundamental rules of modern evolutionary biology: through random mutations and the competition among individuals for reproductive success. One has to wonder: If it is so easy for cheaters to triumph, how can cooperation survive at all?

STRENGTH IN NUMBERS

In the 1950s, some scientists explained cooperation in animals with an idea that came to be known as group selection. They argued that a large group of unrelated animals could outcompete another group, just as individuals outcompete other individuals. The adaptations that allow some groups to outreproduce other groups should become more common over time. Group selection could produce traits and behaviors that benefit the many, not the few. In some bird colonies, for example, only a third of the adults might reproduce in a year. Group selectionists argued that the birds are restraining themselves so that the colony will not get too big and destroy its food supply. They even saw death as resulting from group selection, clearing away old individuals so that young ones can get enough food to reproduce.

Group selection was popular for a time. People began to speak of behavior that was for the good of the species. But by the 1960s, critics were beginning to demolish the theory. They pointed out that group selection can produce benefits only slowly—far more slowly than the changes created by natural selection acting on individuals, as with the rise of cheaters. George Williams, an evolutionary biologist at the State University of New York, Stony Brook, distilled many of these arguments into a devastating assault. In his 1966 book, *Adaptation and Natural Selection,* Williams argued that the group-selection arguments were often the result of lazy thinking. If scientists couldn't see how natural selection produced an adaptation, it was likely they had simply failed to think seriously enough about the question.

Williams declared that most aspects of biology, no matter how puzzling, were the result of strict natural selection working on individuals. Take the school of fish swimming like a superorganism. It might seem as if every individual cooperates for the good of the group, working with others to avoid predators, even if that means an individual gets devoured in the process. Williams argued that the schooling behavior could emerge as each fish tries to boost its personal chances of survival, either by trying to get in the middle of the school or by watching other fish for signs of approaching predators.

Meanwhile, in England, another young biologist, William Hamilton, realized that something important had been ignored in the debate over natural selection and group selection: family. Natural selection favors mutations that spread genes through a population, and one way to spread those genes is by having a lot of healthy children. Hamilton demonstrated, however, that an individual can spread its genes by helping its relatives breed.

Hamilton made his point mainly with social insects, such as ants and bees. A sterile female worker ant may have no hope of reproducing, but that does not mean the genes she carries have no chance of getting into the next generation. Every female worker in an anthill is the offspring of the queen, as are the eggs she helps to raise. That means the worker is helping to rear ants that share some of the same genes she carries. In fact, thanks to a quirk in insect genetics, a worker ant shares more genes with the eggs of the queen than she would with her own offspring. Hamilton put together a mathematical model of genes passing from one generation to the next. If altruism is more likely to pass a set of genes to the next generation than is reproducing oneself, it could be favored by natural selection. Group selection is indeed possible, Hamilton argued, if the group is an extended family.

Williams and Hamilton had a staggering impact on biology. It's as if they had passed out decoder rings that allowed scientists to decipher many mysterious patterns in nature—why some animals dote on their offspring while others abandon them at birth, for example. They could make predictions about the intimate details of species with mathematical precision. As zoologists, Williams and Hamilton didn't have much to say about the evolution of a microbe such as *E. coli.* But it turns out that in many ways *E. coli* supports their view of life as well.

There may be nothing terribly mysterious, for example, about why cheating *E. coli* haven't completely taken over. Cheaters can exploit their fellow bacteria in the stationary phase, but only at a cost. The mutation that turns ordinary *E. coli* into cheaters occurs on a gene called rpoS. Normally rpoS acts as a master control gene, responding to signs of stress by turning on hundreds of other genes. Starvation and other kinds of stress cause rpoS to put *E. coli* into the stationary phase. If a mutation disables rpoS, the microbe will not shut down its metabolism but instead will begin to feed and grow.

Like many other genes, rpoS has many roles to play in *E. coli*'s life. When the microbe enters our stomachs and senses that it has entered an acid bath, rpoS responds by switching on acid-resistance genes. Cheaters cannot marshal these defenses, and so they are more likely to die before they can pass through the stomach. Although cheaters may thrive in one state, they lose out over *E. coli*'s entire life cycle.

Even *E. coli*'s biofilms, those lovely cooperative ventures built on self-sacrifice, may not be quite the models of altruism they seem to be. Joao Xavier and Kevin Foster, two biologists at Harvard, have found evidence that conflict can help produce biofilms. Xavier and Foster built a complex mathematical model of a biofilm to compare how well two kinds of bacteria would fare: one kind produced a biofilm glue (technically known as extracellular polymeric substances), and the other did not. Xavier and Foster seeded an empty surface with both kinds of bacteria and let them grow by eating glucose and consuming oxygen.

At first, Xavier and Foster found, the glue-making bacteria lost out to the others because they were diverting energy from their growth. But soon the balance tipped. As the bacteria multiplied, they used up the oxygen around them and could not grow as quickly. The glue makers could escape suffocation because they created a rising mound on which their offspring could grow. They could reach higher concentrations of oxygen, which allowed them to grow faster, which in turn allowed them to build their mounds of glue even higher. As a mound grew, it suffocated the older bacteria underneath while their descendants—and thus their genes—lived on. Meanwhile, the bacteria that did not make glue, trapped in the biofilm with no way to escape, were buried by their competitors. In some ways a biofilm may be less like a city and more like a forest, in which trees become wooden towers in order to reach the sunlight and avoid the shade of their rivals.

Conflict and cooperation strike an uneasy balance whenever many cells come together, whether they are *E. coli* or the cells of our own bodies. We descend from single-celled ancestors that probably looked a lot like amoebas. At some point our ancestors began to form colonies, which gradually evolved into vast collectives, otherwise known as animals. They communicated with one another as they had before, but now their signals caused them to differentiate into different types of cells, forming tissues and organs. Each time a new animal took shape, most of the cells of its body had to make the ultimate evolutionary sacrifice. They would become part of the body, and in that body they would die. Only sperm and eggs had the slightest chance of their genes surviving.

This is not a simple way to exist. In order to form a full-grown body, most cells must divide many times and then stop. Some kinds of cells must not lose the ability to regenerate themselves, but they must multiply only as much as necessary to heal a wound or build a new intestinal lining. Unfortunately, a dividing cell can mutate, just like a dividing *E. coli.* In some cases, the mutations will turn the cell into a rebel. It will reproduce rapidly, ignoring the signals that tell it to stop. It will produce a mass of insurgent cells, and within that mass new mutations will arise, producing even more rebellious cells. They will develop new tricks for evading the body's defenses and for manipulating the body so that it brings them new blood vessels in order to supply them with extra oxygen and nutrients. They become cheaters, just like the cheaters that exploit *E. coli*'s cooperation. We call their success cancer.

PLACE YOUR BETS

When a starving colony of *E. coli* gets a supply of lactose, there is only one good decision to make: start manufacturing beta-galactosidase and use it to break the lactose down. Some microbes will make the right choice while others will not. The losers keep their lactose-digesting genes shut off, and they continue to starve.

These microbes are all genetically identical, which means that the same genetic circuitry gives rise to both decisions. If natural selection favors genes that boost the reproduction of *E. coli,* how could it have produced this sort of confusion? That is a difficult question, one scientists have only begun to take seriously. The answer they have now settled on is this: *E. coli* is a smart gambler.

Every gambler who comes to a racetrack hopes to place a winning bet. The best way to win is to see into the future and know which horse will come in first. But in the real world, gamblers can only hope to winnow down their choices to a few strong horses. Even with this limited selection, they run the risk of losing money. Some gamblers shield themselves against losses by hedging their bets. They wager on several horses in the same race. If one of their horses wins, they get money. They don't leave the racetrack with as much money as they would have had they bet only on the winner. But the other bets can act as a good insurance policy. If one of the other horses wins, the gambler can still go home with more money than he came with.

Gamblers aren't the only people who think about hedging bets. Mathematicians and economists have explored all sorts of variations on the strategy, many of which have been borrowed by stockbrokers, bankers, and other people who make uncertain choices about how financial markets behave. A broker buying stock in a biotechnology company may sell short on a different one, so that no matter which way the market goes she makes money. And evolutionary biologists have now borrowed the mathematics of the marketplace to understand why *E. coli* clones can act so differently.

E. coli's gamble consists of choosing a response to a particular situation. In some cases, the choice is clear. A population of microbes should all respond in the same way. But in other cases, it pays for the population to hedge its bets. In other words, it pays for some individuals to respond one way and others to respond in another.

Which way *E. coli* should bet depends on how much information it can get. If *E. coli* can get a lot of reliable information, it makes sense to put all its money on one bet. But in other cases, it may be hard to determine the best choice. Conditions may be changing quickly and unpredictably. *E. coli* may be better off hedging its bet in these cases, allowing individuals to respond in different ways.

Lactose, for example, causes *E. coli* to hedge its bets. A supply of lactose may allow bacteria to survive when other kinds of sugar have been devoured. But in order to feed on lactose, a microbe first has to clear away all the proteins it was using to feed on other sugars and begin making the proteins it needs to eat lactose. That's a lot of time and energy for a microbe to invest. The investment may pay off or it may be a waste of effort—the lactose may disappear quickly or a more energy-rich sugar may turn up.

E. coli hedges its bets by using its unpredictable bursts of proteins to create both eager and reluctant individuals. If the colony happens to encounter some lactose, the eager microbes will be ready to seize the moment, while the others respond more slowly. If the surge of lactose never comes, the reluctant microbes will grow faster because they haven't wasted energy preparing for a feast that never arrived. Either way, the colony benefits.

E. coli hedges many bets, scientists are finding, and some of those bets make us sick. Strains of *E. coli* that infect the bladder need to make sticky hairs to attach to host cells, but the hairs draw the attention of the immune system. To balance this trade-off, the bacteria hedge their bets by randomly switching on and off the machinery for making the hairs. At any moment some individuals in a colony will sprout hairs and others will remain bald.

Bet hedging may also help *E. coli* defend itself against antibiotics. Many antibiotics kill *E. coli* by attacking the proteins the microbes use to grow. When antibiotics encounter a population of susceptible *E. coli,* they kill it off with staggering swiftness. Or at least they kill most of the microbes off. About 1 percent of the *E. coli* in a biofilm can survive an attack of antibiotics for hours or days. The survivors can rebuild the biofilm and make a person as sick as before.

This resilient minority carries no special genes for resisting antibiotics. They are genetically identical to their dead relatives. Scientists can isolate the survivors and allow them to multiply to form large colonies. The new colonies will be just as vulnerable to antibiotics. Once again, about 99 percent of the microbes die and 1 percent persist.

Scientists discovered so-called persister bacteria in 1944, and for the next sixty years they remained almost entirely baffled by them. Some researchers suggested that antibiotics drive a few microbes into a mysterious dormant state in which they can escape damage. A team of scientists led by Nathalie Balaban of the Hebrew University of Jerusalem tested this idea in 2004 by building a device to spy on persister cells. The scientists placed *E. coli* in microscopic grooves just wide enough to hold a single microbe. When an individual *E. coli* divided, its offspring remained in a neat line. Balaban and her team could watch the lines stretch and measure how quickly lineages of cells grew.

After several generations, the scientists doused the bacteria with a potent antibiotic. Most of the *E. coli* died, but the persisters remained. Balaban and her colleagues found that the persisters grew far more slowly than normal cells, although they had not stopped growing altogether. By looking back at their earlier measurements, Balaban discovered that the microbes had become slow-growing persister cells *before* the antibiotics arrived.

Balaban concluded that every *E. coli* has a tiny chance at any moment of spontaneously turning into a persister. Once it makes the change, the microbe has a small chance of reverting to a normal fast-growing cell. All the bacteria Balaban studied, persisters and growers alike, were genetically identical, which meant that mutations were not the source of persistence. Yet persisters gave rise to more persisters, as if persisting were a hereditary trait.

Persisters are born of noise. That's the theory of Kim Lewis, a leading expert on the phenomenon at Northeastern University. Lewis and his colleagues have found a way to compare the proteins produced by persister cells with the ones made by normal *E. coli.* One of the major differences between the two kinds of bacteria is that persister cells make a lot of toxins. Scientists have long puzzled over these toxins, which lock on to *E. coli*'s other proteins and stop them from doing their normal jobs. In most of the bacteria, these toxins don't cause any harm because *E. coli* also produces their antidotes—antitoxins that grab the toxins before they can interfere with the microbe's physiology.

It's these toxins, Lewis argues, that turn *E. coli* into persisters. Normally *E. coli* churns out a tiny stream of toxins, along with another tiny stream of antitoxins. But thanks to the noisy workings of its genes, the microbe sometimes hiccups, releasing a burst of toxins. The extra toxins that aren't disabled by *E. coli*'s small supply of antitoxins are free to attack proteins. They don't do any permanent damage, but they do bring *E. coli*'s growth nearly to a halt. After the outburst, *E. coli*'s toxins gradually dwindle as *E. coli* produces more antitoxins. Once its proteins are liberated, it can go back to being a normal microbe again.

This noisy network acts like a roulette wheel, randomly picking out a few individuals to stop growing at any moment. It's usually a bad thing for an individual microbe to get stuck with extra toxins, because a persister will fall behind the other, fast-growing *E. coli*. But there's also a small chance that a disaster will strike while the microbe is a persister. That disaster might come in the form of a pencillin pill, or it might be a naturally produced poison released by another microbe. In either case, the persister will turn out to be the big winner. For the entire population of *E. coli*, it doesn't matter which individual wins as long as its individuality-generating genes continue to get passed down to new generations.

SPITEFUL SUICIDE

Persister cells make a sacrifice for their companions, giving up the chance to multiply quickly. But when *E. coli* produce colicins, the chemical weapons for killing rival strains, they pay a far bigger price. In order to let their kin thrive, they explode in a suicidal blast.

The chemical warfare practiced by *E. coli* is the dark side of altruism. William Hamilton originally argued that natural selection could favor sacrifice if it meant an individual could help its relatives reproduce more. In 1970, he recognized that natural selection could also favor sacrifice if it meant that nonrelatives suffered—a nasty sort of altruism he called spite. Hamilton always argued that spite was rare and inconsequential, because his equations suggested it would be favored only when populations were very small. But in 2004, Andy Gardner and Stuart West at the University of Edinburgh demonstrated that if unrelated individuals compete fiercely with their immediate neighbors spite can also evolve.

E. coli meets these spiteful standards. It competes in the crowded confines of the intestines for a limited supply of sugar. An individual microbe sacrifices its own reproductive future by committing suicide, but its colicins destroy many competitors, allowing the microbe's own close relatives to thrive. As with persistence, becoming a colicin maker is a matter of chance. The noisy production of proteins determines which few individuals will respond to starvation by switching on their colicin-producing genes. The burden is shared by all.

Spite, some experiments now suggest, may also drive *E. coli* to become more diverse. Margaret Riley, a biologist at the University of Massachusetts, Amherst, and her colleagues have observed the evolution of this arms race in experiments on *E. coli* in both petri dishes and the guts of lab mice. Once in a rare while, an antidote gene may mutate into a more powerful form. Instead of just defending *E. coli* against its own colicin, it can also defend against the colicins made by other strains. This mutation gives a microbe an evolutionary edge, because it can survive enemy attacks that kill other members of its strain.

This powerful antidote opens the way for another advance. A second mutation strikes the colicin-producing gene, causing it to make a new colicin. Its relatives, which still carry an antidote for the old colicin, are killed off by the mutant toxin. But thanks to its powerful antidote, the microbe that makes the new colicin can survive while its relatives die. Its spite becomes intimate.

The emergence of new colicins drives the evolution of new antidotes in other strains. Likewise, new antidotes drive the evolution of new colicins. But *E. coli* has to pay a price for this weaponry. It has to use energy to make colicins and antidotes, which are particularly big as bacterial molecules go. A new colicin may be even deadlier than its predecessor, but it may also become a drain on a microbe. If a mutation leaves a microbe unable to make colicins—but still able to resist them—it may be able to channel the extra energy into reproducing. A colicin-free strain will spread, outcompeting the colicin makers.

If colicin makers are driven to extinction, their colicins no longer pose a threat to neighboring bacteria. Now antidotes become a waste of effort, since there is nothing for them to protect *E. coli* against. Natural selection can begin to favor pacifists—microbes that make neither colicins nor antidotes. Once the pacifists come to dominate the population, colicin producers can invade the population once more, killing off the vulnerable strains and getting the food for themselves. And so the journey comes full circle.

These sorts of cycles emerge spontaneously from evolution. You can think of them as games in which players use different kinds of strategies to compete with other players. In the case of *E. coli,* a strategy might be to make a particular colicin or to do without colicins and antidotes altogether. In the case of a male elephant seal, strategies might include fighting with other males for the opportunity to mate with females or sneaking off with a female without the big male on the beach noticing. In some cases, one strategy may prove superior to all the others. In other cases, two strategies may coexist. Fighting males and sneaker males can coexist in many species, for example. In still other cases, the success of strategies goes up and down over time.

Scientists sometimes call this cycling evolution a rock-scissors-paper game. In the game, each player can make a fist for a rock, extend two fingers for scissors, or hold the hand flat for paper. A player wins or loses depending on what the other players do. Rock beats scissors, but scissors beats paper, and paper beats rock. If a population of organisms is dominated by one strategy—call it paper—then natural selection will favor scissors. But once scissors takes over, rock is favored, then paper, and so on.

The common side-blotched lizard of coastal California plays a colorful version of rock-scissors-paper. The male lizards have colored throats, which may be orange, yellow, or blue. The orange-throated lizards are big fighters; they establish large territories with several females. The blue-throated lizards are medium sized; they defend small territories, holding just a few females, which they can guard carefully. The yellow-throated males are small and sneak around for mates, taking advantage of the fact that they look like females. Each type of male can outcompete one type but not the other. The yellow-throated males can sneak past the orange-throated males because the territories of the orange-throated males are so big. The yellow-throated males cannot use the same strategy against the blue-throated males because the blue-throated males stay close to their females and are bigger than the yellow-throated males. But the blue-throated males lose against the orange-throated males because the orange-throated males are bigger.

Over a period of six years, each type of male goes through a population cycle. When the orange-throated males become common, natural selection favors yellow-throated males, which can sneak off with their females. But once yellow-throated males become common, the biggest benefits go to blue-throated males, which can fight off the yellow-throated males and father lots of baby lizards with their few females. And in time, natural selection favors the orange-throated males again.

When scientists at Stanford and Yale discovered the *E. coli* version of the rock-scissors-paper game in 2003, they suggested that it may turn out to be particularly common. Chemical warfare is a frequent strategy in nature, particularly among organisms that are too small or too immobile to use other sorts of weapons. Trees poison their insect visitors, corals ward off grazers, and humans and other animals produce antibodies to fight off pathogens. The race to develop better poisons and defenses, as well as the added dimension of the rock-scissors-paper game, can foster the evolution of diversity. Scientists have long known that a single strain of *E. coli* may dominate the gut for a few months, only to later shrink away, making way for a rarer strain. The colicin war may be one force behind this cycle.

E. coli may be able to spontaneously evolve a harmonious food web. But when it comes to weaving Darwin's tangled bank, war may be just as good as peace.

DEATH COMES TO ALL

Not long ago, *E. coli* was immortal. That's not to say it was invulnerable. The bacteria can die in all sorts of ways—devoured by protozoans, starved for years in a famine, or ripped open like a water balloon by the prick of a colicin needle. But decades of gazing at *E. coli* left scientists convinced that death is not inevitable. Left to its own devices, *E. coli* remained eternally young. Here was one way, at least, in which *E. coli* was fundamentally different from us. Our bodies slide into decay on a relatively tight schedule. Our immune system lets more viruses and bacteria invade our bodies unchallenged. Our brains shrink; our bones grow brittle; our skin droops.

The question of why we slide this way toward death preoccupied George Williams. He was so fascinated by it that he charted his own decline. Beginning at age fifty-two, he would go once a year to a track near his home on Long Island and time how long it took him to run 1,700 meters. Some years he ran a little faster than he had the previous year, but over the course of twelve years he gradually slowed down. Why, Williams wanted to know, was he declining so steadily? If he had to die, why couldn't he stay young and fit until his body suddenly gave out? And if he did have to get old, why did his decline follow the particular downward curve that it did? Why hadn't he run so slowly in his twenties instead of in his fifties?

After all, Williams could look to the natural world for an endless supply of alternatives. A clam may live for four centuries. At the other extreme are salmon, which return in peak condition to the streams where they were born. They find a mate, have baby salmon, then promptly grow old at catastrophic speed and die. In a few weeks the salmon age more than humans do in a few decades.

As a graduate student in the 1950s, Williams had listened to his teachers explain that death existed for the good of the species. The old had to make way for the young, or else a species would become extinct. Williams thought that was nonsense. Instead, he considered how natural selection acting on individuals might create old age. Williams argued that it could be a side effect caused by genes that offered advantages in youth. As long as the advantages of these genes outweighed the disadvantages, they would become widespread. Cancer, declining stamina, deteriorating vision, and the other burdens of old age might all be the result of natural selection.

Williams argued that organisms face these sorts of evolutionary tradeoffs throughout their lifetime. How much energy should they invest in maturing before they start to have babies, for example, or how much energy should they invest in raising offspring before they search for another mate? Natural selection ought to find the balance between those demands. Williams speculated that animals could also keep track of how those factors change over their lifetime and adjust their behavior accordingly, like an investor deciding which stocks to keep or sell.

Over the past forty years, Williams's theory has evolved into an experimental science of aging. Now scientists can predict which species will get old and why. In a 2005 study, to pick just one example from hundreds, scientists studied the sockeye salmon that return to Pick Creek, Alaska, each year. The salmon come back in July and August. Once the female salmon have mated, they select a spot to lay their eggs and dig a nest in the gravel bottom of the creek. After they lay their eggs, they cover them and guard them from other females that might want to take over the nest to lay their own eggs.

The salmon of Pick Creek face just the sort of trade-off Williams proposed. Once they leave the ocean to travel to their breeding grounds, they stop eating for good. They have only a fixed amount of energy to divvy up among the things they do before they die. The females have to put some of their energy into their developing reproductive system in order to make eggs. They can also put some of their energy into maintaining their bodies so that they will live long enough to fight off other salmon. It's a zero-sum game.

The scientists predicted that salmon arriving at Pick Creek earlier in the season would live longer than the salmon that came later. A salmon that lays its eggs in July has weeks of battling ahead. If it puts all its energy into eggs and dies early, other salmon will take over its nest, and its genes won't have a chance of getting into the next generation. If a late-arriving salmon invests its energy in long life, it's wasting its effort, since it will still be alive when the rest of the salmon have died off. Late arrivers should invest in making extra eggs.

When the researchers compared early and late arrivers, they found their predictions met. The early arrivers survived on average for twenty-six days at Pick Creek, whereas the late arrivers survived only twelve days. The early arrivers put roughly an equal amount of energy into maintaining their bodies and protecting their eggs. The late arrivers put twice as much energy into protecting their eggs as into maintaining their bodies.

Williams's predictions work not just for salmon but for fruit flies, vinegar worms, guppies, swans, humans, and many other species as well. But until recently experts on aging considered *E. coli* off limits. A trade-off between long life and reproduction seemed simply not to exist. *E. coli* did not have parents and children. An individual *E. coli* just duplicated its DNA and pulled itself apart into two new individuals. The parent became the children. Starvation might slow *E. coli* down, and chemical warfare and other assaults might kill the bacteria outright. But left to themselves with enough food, *E. coli* would reproduce forever, each new microbe as healthy as its forerunners.

That was what scientists thought until Eric Stewart, a microbiologist now at Northeastern University, decided to take a very close look at *E. coli.* He and his colleagues built a sort of voyeuristic *E. coli* paradise. They injected a single microbe onto an agar-coated slide, covering the little shelter with glass and sealing the sides shut with silicon grease. The microbe carried a light-producing gene, making it easy to film through the top of the slide. The scientists mounted the slide on a microscope, and the entire apparatus was put in a box that was kept as warm as a healthy human gut.

The single *E. coli* feasted and divided. Its descendants spread out in a layer one cell deep. At regular intervals a camera mounted on the microscope took a picture of the glowing colony. Comparing one picture with the next, Stewart could track the fate of every branch of his *E. coli* dynasty. He could time how long it took each microbe to divide and then how long its two offspring needed and then its four grandchildren. Given that all the microbes were genetically identical and all were living in the same perfect conditions for growth, they all should have grown at the same rate. But they didn't. Some individuals grew more slowly than their siblings, and over time their descendants lagged farther and farther behind.

Some bacteria, Stewart discovered, were getting old. Each time a microbe reproduces, it builds itself a ring in order to cut itself in half. At the same time, it builds two new caps to cover the new ends of its daughter cells. When those two daughter cells split, each will build new caps as well. After several generations, some bacteria will have old poles and others will have new ones. In the diagram below, the numbers show how many generations have passed since a cap has been created:

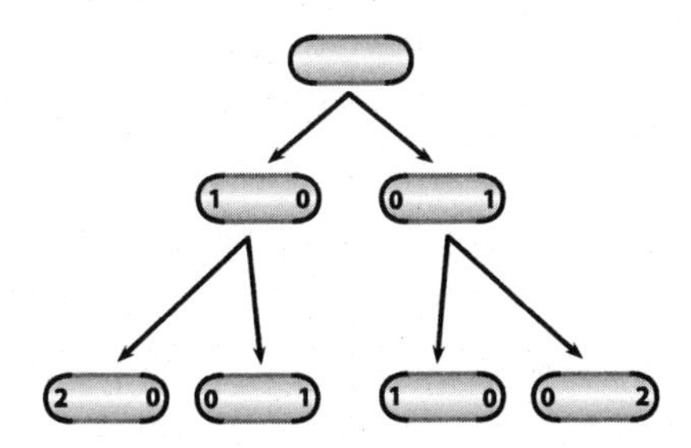

Stewart discovered that as the caps on microbes got older, the microbes grew more slowly. He estimates that the aging *E. coli* were slowing down so quickly that after a hundred generations they would stop dividing altogether.

Once more the Williams-Hamilton decoder ring can help. Old age must have some evolutionary advantage over immortality for *E. coli.* Its edge may come from the inescapable damage that strikes the bacteria. Proteins become snarled; genes mutate. When a microbe divides, it may pass down its defective proteins and genes to one or both of its descendants. Over the generations, more and more damage can pile up like a cruel, compounding legacy. Of course, *E. coli* can fix this damage, and it does fix a lot of it. Yet that repair doesn't come free. A microbe must use up a lot of energy and nutrients to repair itself. If it spent all its resources on repair, a more careless microbe would outcompete it.

There is another way to cope with damage: push it all into one place. In *E. coli*'s case, the dumping grounds are its poles. *E. coli* does not put much effort into repairing them, and when it divides, each of its descendants gets an old, damaged pole along with a new one on the other end. Over the generations, some of the poles can get very old—and presumably accumulate a lot of damage to their proteins. Instead of trying to be a perfectionist, Stewart suggests, *E. coli* may just turn its poles into garbage cans. A microbe that lets some of its descendants get old while the rest stay young may have found the best strategy for evolutionary success.

What once seemed like a major exception to Monod's rule has now vanished. Once again *E. coli* has hit on the same strategy we humans have. When a fertilized human egg begins to grow into an embryo, it soon develops into two types of cells: cells that can become new people (eggs and sperm) and all the others. We invest a great deal of energy in protecting eggs and sperm from the ravages of time and much less on protecting the rest of our bodies. From this unconscious choice, we allow our progeny to live on while we die. For both humans and *E. coli,* the privilege of life must be paid with death.

Seven

DARWIN AT THE DRUGSTORE

LIFE AGAINST LIFE

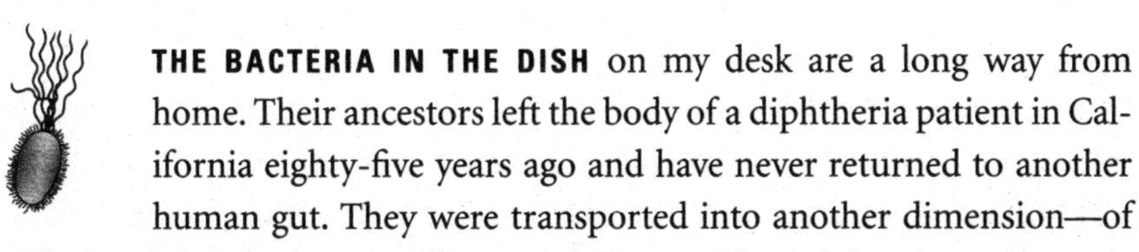

THE BACTERIA IN THE DISH on my desk are a long way from home. Their ancestors left the body of a diphtheria patient in California eighty-five years ago and have never returned to another human gut. They were transported into another dimension—of flasks and freezers, centrifuges and X-rays. These laboratory creatures have enjoyed a strange comfort, gorging themselves on amino acids and sugar. And over hundreds of thousands of generations they have evolved. They have become fast breeders and have lost the ability to survive for long in the human gut. They avoid extinction only because they have become so dear to the biologists who carry them from flask to freezer to incubator.

Over those eighty-five years their wild cousins have gone on with their own lives. They have continued to colonize guts, and they have evolved as well. The microbes that live inside us today are not the same as the ones that lived inside people in 1920. We are the source of much of that change.

The most obvious way we have changed *E. coli* is by trying to fight infections with drugs. *E. coli* and other bacteria have responded to those drugs with a rapid burst of evolution. They can now resist drugs that once would have wiped them out. Scientists are now left scrambling to find new drugs to replace the failed ones, and there's little reason to think *E. coli* and other microbes won't evolve resistance to them as well.

While some scientists have observed *E. coli* evolve in their laboratories, we have also launched a global, unplanned experiment in *E. coli* evolution. Like laboratory experiments, the rise of resistant *E. coli* is offering its own clues to the workings of evolution. Resistance can evolve through the familiar course of random mutations and natural selection. But in some ways, *E. coli* is not fitting into the conventional picture. In the evolution of resistant *E. coli,* some researchers claim to have found evidence that the microbe can alter the way it mutates to suit the conditions it faces. And while Darwin erected his theory on the idea that organisms inherit traits from their direct ancestors, *E. coli* has acquired much of its resistance to antibiotics from other species of bacteria, which can trade genes like business cards. These discoveries are significant not only because they may help in the battle against drug-resistant pathogens. They may also reveal forces that have been shaping life for the past 4 billion years.

The era of antibiotics began suddenly, but it followed a long, slow prelude. Traditional healers long knew that mold could heal wounds. In 1877, Louis Pasteur found that he could halt the spread of anthrax-causing bacteria by introducing "common bacteria" in their midst. No one knew what the common bacteria did to stop the anthrax, but scientists gave it a name anyway: *antibiosis,* the ability of one creature to kill another.

In 1928, Alexander Fleming, a Scottish bacteriologist, discovered a molecule that could kill bacteria. He noticed that one of his petri dishes had become contaminated with mold. There were no bacteria near it. He ran tests on the mold and discovered that it could halt the spread of bacteria. Yet it did not harm human cells. Fleming isolated the mold's antibiotic and named it penicillin.

At first, penicillin did not look like a promising drug. For one thing, Fleming could extract only tiny amounts of it from mold, and it proved too fragile to be stored for very long. It took ten years for penicillin to live up to its promise. Howard Florey and Ernst Chain at Oxford University figured out how to coax the mold to make enough penicillin to test on mice. They infected mice with streptococci and injected some with penicillin. The treated mice all survived, and the others all died. In 1941, Florey and Chain persuaded American pharmaceutical companies to adopt their penicillin production scheme and expand it to an industrial scale. By 1944, wounded Allied soldiers were being cured of infections that would have killed them a year before. In the next few years, a rush of other antibiotics came along, mostly derived from fungi and bacteria.

Antibiotics, scientists discovered, kill bacteria in many ways. Some attack enzymes that help replicate DNA. Others, such as penicillin, interfere with the construction of the peptidoglycan mesh that wraps around *E. coli* and other bacteria. Gaps in the mesh form, and the high-pressure innards of the microbes burst out. Organisms naturally make only trace amounts of antibiotics, but drug companies began to produce them in enormous bulk, rearing fungi and bacteria in giant fermenters or synthesizing drugs from scratch. It would take billions of microbes to produce the antibiotics in a single pill. In such a concentrated form, antibiotics had a staggering effect on disease-causing bacteria. They didn't just reduce infections. They got rid of them altogether, and with few noticeable side effects. The war against infectious diseases seemed to have suddenly become a rout.

But even in those heady days of early victory, there were signs of trouble. At one point in their research, Florey and Chain discovered that their cultures of mold had been invaded by *E. coli.* The bacteria were able to survive in a soup of penicillin by producing an enzyme that could cut the antibiotic molecule into feeble fragments.

As penicillin was being introduced to the world, microbiologists were discovering how mutations arose in *E. coli.* In 1943, Delbrück and Luria showed that mutations spontaneously made *E. coli* resistant to viruses. In 1948, the Yugoslavian-born geneticist Milislav Demerec showed that the same held true for antibiotics. He bred resistant strains of *E. coli* and *Staphylococcus aureus.* Both species became increasingly resistant as they picked up a series of mutations. In the same year that Demerec published his results, doctors reported that penicillin was beginning to fail in their *Staphylococcus*-infected patients.

These disturbing discoveries did nothing to halt the rise of antibiotics. Today the world consumes more than ten thousand tons of antibiotics a year. Some of those drugs save lives, but a lot of them are wasted. Two-thirds of all the prescriptions that doctors hand out for antibiotics are useless. Antibiotics can't kill viruses, for instance. Many farmers today practically drown their animals with antibiotics because the drugs somehow make the animals grow bigger. But the cost of the antibiotics is greater than the profit from the extra meat.

Along with the rise in antibiotics has come a rise in antibiotic resistance. Drugs that were once fatal to bacteria are now useless. *E. coli*'s story is typical. Resistant strains of *E. coli* first emerged in the 1950s. At first only a small fraction of *E. coli* could withstand any particular antibiotic, but over several years resistant microbes became more common. Soon the majority of *E. coli* could withstand the drug. As one drug faltered, doctors would prescribe another—a stronger drug with harsher side effects or a more recently discovered molecule. And in a few years that drug would begin to fail as well. Before long, strains of *E. coli* emerged that could resist many antibiotics at once.

E. coli uses many tricks to dodge antibiotics. As Florey and Chain discovered, it can secrete enzymes that cut penicillin into harmless fragments. In some cases, *E. coli*'s proteins have taken on new shapes that make it difficult for antibiotics to grab them. And in other cases, *E. coli* uses special pumps to hurl antibiotics out of its interior. For every magic bullet science has found for *E. coli, E. coli* has acquired an equally magic shield.

SKIN OF FROG

E. coli has evolved its resistance to antibiotics almost entirely out of view. It was not trapped in a laboratory flask, where a scientist could track every mutation from one generation to the next. Its flask was the world.

The pieces of evidence scientists have assembled are enough for them to reconstruct some of its history. The genes that now provide *E. coli* with resistance to antibiotics did not suddenly appear in 1950. They descend from older genes that originally had other functions. Some of the pumps that flush antibiotics out of *E. coli* probably evolved from pumps bacteria use to release signaling molecules. Others originally flushed out the bile salts *E. coli* encounters in our guts.

When *E. coli* first encountered antibiotics, its pumps probably did a poor job of getting rid of them. But on rare occasion the genes for the pumps mutated. A mutant microbe might pump out antibiotics a little faster than others. Before modern medicine, such mutants wouldn't have been any better at reproducing than other bacteria. Their mutations might even have been downright harmful. But once they began to face antibiotics on a regular basis, the mutants had an evolutionary edge.

That edge may have been razor thin at first. Only a few of the resistant mutants might have survived a dose of antibiotics, but that was better than getting exterminated. Over time, resistant mutants became more common in populations of *E. coli*. Their descendants acquired new mutations that made them even more resistant. In 1986, scientists discovered strains of *E. coli* that made an enzyme able to destroy a group of antibiotics called aminoglycosides. In 2003, another team discovered *E. coli* carrying a new version of the gene. It had two new mutations that made it resistant not just to aminoglycosides but also to a completely different antibiotic, called ciprofloxacin.

Even within a single person, *E. coli* can evolve to dangerous extremes. In August 1990, a nineteen-month-old girl was admitted to an Atlanta hospital with a fever. Doctors discovered that *E. coli* had infected her blood, probably through an ulcer in her intestines. Tests on the bacteria revealed that they were already resistant to two common antibiotics, ampicillin and cephalosporin. Her doctors gave her other antibiotics, each more potent than the last. Instead of wiping out her *E. coli,* however, they made it stronger. It acquired new resistance genes, and the ones it already had continued to evolve. After five months and ten different antibiotics, the child died.

Terrifying failures like this one leave scientists hoping that someday they will find new antibiotics that are immune to the evolution of resistance. Like Fleming before them, they find promising new candidates in unexpected places. One particularly promising group of molecules was discovered in 1987 in the skin of a frog.

Michael Zasloff, then a research scientist at the National Institutes of Health, noticed that the frogs he was studying were remarkably resistant to infection. At the time, Zasloff was using frogs' eggs to study how cells use genes to make proteins. He would cut open African clawed frogs, remove their eggs, stitch them back up, and put them in a tank. Sometimes the water in the tank became murky and putrid, yet his frogs—even with their fresh wounds—did not become infected.

Zasloff suspected the frogs were making some kind of antibiotic. He ground up frog skin for months until he isolated a strange bacteria-killing molecule. It was a short chain of amino acids known as a peptide. He and other researchers discovered that it is fundamentally different from all previously discovered antibiotics. It has a negative charge, which attracts it to the positively charged membranes of bacteria but not to the cells of eukaryotes such as humans. Once the peptide makes contact with the bacteria, it punches a hole in their membranes, allowing their innards to burst out.

Zasloff realized he had stumbled across a huge natural pharmacy.

Antimicrobial peptides, it turned out, are made by animals ranging from insects to sharks to humans, and each species may make many kinds. We produce antimicrobial peptides on our skin and in the lining of our guts and lungs. If we lose the ability to make them, we become dangerously vulnerable. Cystic fibrosis may be due in part to mutations that disable genes for antimicrobial peptides produced in the lungs. The lungs become loaded with bacteria and swell with fluid.

Having discovered antimicrobial peptides, Zasloff now tried to turn them into drugs. They might be able to wipe out bacteria that had evolved resistance to conventional antibiotics. Antimicrobial peptides might even be resistance proof. In order to become resistant to antimicrobial peptides, bacteria would have to change the way they build their membranes. It was hard to imagine how bacteria could make so fundamental a change in their biology, and experiments seemed to back up this hunch. Some scientists randomly mutated *E. coli* to see whether it could produce mutants able to survive a dose of antimicrobial peptides. No luck.

But an evolutionary biologist named Graham Bell at McGill University in Montreal suspected that *E. coli*—and its evolutionary potential—might be more powerful than others had thought. Michael Zasloff, for one, didn't think so. But as a good scientist, he was willing to put his hypothesis to the test. He teamed up with Bell and Bell's student Gabriel Perron to run an experiment. Remarkably, his hypothesis failed.

The researchers began by exposing *E. coli* to very low levels of an antimicrobial peptide. A few microbes survived, which the scientists used to start a new colony. They then exposed the descendants of the survivors to a slightly higher concentration of the antimicrobial peptide. Again most of the bacteria died, and they repeated the cycle, raising the concentration of the drug even higher. *E. coli* turned out to have a remarkable capacity to evolve. After only six hundred generations, thirty out of thirty-two colonies had done the impossible: they had become resistant to a full dose of antimicrobial peptides. These results raise some serious concerns about how effective antimicrobial peptides will be when they hit the market. *E. coli* and other bacteria that are hit by low doses of antimicrobial peptides may evolve resistance. If they do, they will survive stronger and stronger doses until they can withstand the full strength of these drugs.

If *E. coli* can evolve resistance to antimicrobial peptides so quickly, then how did they protect Zasloff's dirty frogs? *E. coli* and other bacteria are locked in an evolutionary race with the animals they colonize. When an animal evolves a new antimicrobial peptide, natural selection favors microbes that can resist it. One common counterstrategy is for a microbe to make an enzyme that can cut the new peptide into pieces before it is able to do any damage.

Now the evolutionary pressure shifts back to the animal. Mutations that allow an animal to block the peptide-cutting enzyme may allow it to survive infections. It will pass down the mutation to its descendants. Animals defend against peptide-slicing enzymes by stiffening the peptides. The peptides are folded over on themselves and linked together with extra bonds. But microbes have evolved counterstrategies of their own. For example, some species secrete proteins that grab the antimicrobial peptides and prevent them from entering the bacteria.

One of the most potent ways for animals to overcome all of these strategies is by making lots of different kinds of antimicrobial peptides. New ones can be produced by gene duplication or by borrowing peptides with other functions. The more antimicrobial peptides an animal makes, the harder it is for bacteria to recognize them all. Thanks to this arms race, the genes for antimicrobial peptides have undergone more evolutionary change than any other group of genes found in all mammals.

Compared with this complex, ever-changing attack on antimicrobial peptides, Bell and Zasloff's experiment is child's play. They exposed *E. coli* to a single kind of antimicrobial peptide and created a strong advantage for mutants that could withstand it. Unfortunately, modern medicine works more like Bell and Zasloff's experiment than like our own evolution. Doctors have only a few antibiotics to choose from when fighting an infection, and they generally prescribe only a single drug to a patient. In a few years this practice gives us resistant bacteria. We might do a better job of fighting bacteria if new drugs came through the development pipeline faster and if doctors could safely prescribe several of them at once.

There are many lessons to be learned from *E. coli*'s quick evolution of resistance. The most surprising of all is that our own bodies, and those of our ancestors, are actually drug-development laboratories.

EVOLUTION ON DEMAND

When Salvador Luria discovered the jackpot pattern in *E. coli*'s resistance to viruses in 1942, he provided some of the first compelling evidence that mutations strike randomly and blindly. A vast number of other experiments on *E. coli* and many other species confirm the steady rate of mutations. But there are a few experiments on *E. coli* that raise some intriguing doubts. Perhaps some mutations are not so blind after all.

Floyd Romesberg, a chemist at Scripps Research Institute in La Jolla, California, carried out an experiment to watch *E. coli* evolve resistance to antibiotics. The drug he chose was ciprofloxacin, or cipro for short. Cipro first emerged in the early 1980s as a promising replacement for older antibiotics that had begun to fail. But within a few years, scattered reports of resistance began to appear. Cipro-resistant bacteria are now very common in some parts of the world. In Germany, 15 percent of *E. coli* were resistant in 2002. In China that same year, one study put the figure at 59 percent.

To understand how cipro-resistant genes evolved, Romesberg and his colleagues injected a disease-causing strain of *E. coli* into six-week-old mice. They then treated the mice with cipro, and the infection disappeared. Or at least it seemed to. Three days later the mice were sick with *E. coli* again. When the scientists tested the bacteria, they discovered that the *E. coli* had become fifty times more resistant to cipro since the start of the experiment.

Cipro kills *E. coli* by tricking it into committing suicide. It interferes with an enzyme known as a topoisomerase, which normally helps to untangle DNA by snipping it and then joining it back together. Once the topoisomerase has cut the DNA, cipro prevents it from finishing its job. The free ends attract other enzymes whose job it is to chop up loose pieces of DNA. They end up destroying much of *E. coli*'s chromosome and thus killing the microbe.

It occurred to Romesberg that cipro might cause *E. coli* to do something else as well: mutate faster. *E. coli* repairs damaged DNA with enzymes called polymerases. It makes two kinds of polymerases: one that does high-fidelity repair and one that does low-fidelity work. The hi-fi polymerases usually handle all the repair work while the genes for lo-fi polymerase are switched off by a protein called LexA. But things change when *E. coli* is in a crisis. When *E. coli* becomes burdened with a lot of damaged DNA, LexA falls off the lo-fi polymerase genes. Now the lo-fi polymerases help repair *E. coli*'s DNA. And because they do a less accurate job, they leave behind more mutations.

Romesberg wondered if these extra mutations helped *E. coli* evolve resistance to cipro faster. While most of the mutations might harm the bacteria, a few might produce topoisomerases that could keep doing their cut-and-paste job even in the presence of cipro. It was possible that extra mutations would arise only during these sorts of crises. Once *E. coli* could cut and paste its DNA again, its supply of loose DNA would dwindle. LexA would grab on to the genes for lo-fi polymerases and shut them down. *E. coli* would return to its more careful DNA repair.

Romesberg and his colleagues tested their hypothesis with an elegant experiment. They engineered a strain of *E. coli* in which LexA did not fall off the lo-fi polymerase genes. Exposed to cipro, these microbes would go on repairing their DNA with exquisite accuracy. Romesberg and his colleagues injected their engineered strain into mice and gave them cipro. In 2005, they reported their results: unable to mutate more, the *E. coli* evolved no resistance to cipro at all.

Romesberg's experiment suggests that *E. coli* is not just passively accumulating mutations. *E. coli* may have evolved ways to manipulate mutations to its own advantage.

The first inklings of not-so-blind mutations came in a "water, water, everywhere, but not a drop to drink" experiment in 1988. John Cairns, then at Harvard, and his colleagues engineered a strain of mutant *E. coli* that was almost completely unable to feed on lactose. Its *lac* operon was in good working order, but the promoter sequence where it could be switched on was slightly mutated. Cairns and his colleagues then gave the bacteria nothing but lactose to eat. The bacteria stopped growing and began to starve. But they did not die out completely.

Over the course of six days, a hundred colonies emerged. Cairns examined their *lac* operon and found that a mutation had struck the microbes, allowing them to switch on the operon again. Cairns calculated that if the bacteria had been mutating spontaneously at their normal rate, only a single colony would have formed in that time. Instead, Cairns concluded, these microbes had acquired working genes a hundred times faster than they should have.

"Cells may have mechanisms for choosing which mutations will occur," Cairns and his co-authors wrote.

These "directed mutations," as they came to be known, caused an uproar. The idea that *E. coli* could respond to a crisis by mutating a specific piece of DNA smacked of Lamarck. Critics claimed that Cairns's hypothesis was practically mystical, requiring *E. coli* to know that mutating a particular part of its DNA would help it in a particular crisis. A wave of other studies followed as scientists tried to figure out just what was happening.

A consensus emerged that these mysterious mutations were not precisely directed toward any particular goal. Many of the bacteria that regained the ability to feed on lactose also carried new mutations on genes that had nothing to do with lactose. Instead of directed mutations, scientists began to speak of "hypermutation." And by *hyper*, they meant that during a crisis *E. coli*'s mutation rates could soar a hundred- or even a thousandfold. Several studies identified *E. coli*'s lo-fi polymerases as the enzymes that created these extra mutations.

Some scientists argue that hypermutation is an elegant strategy to ward off extinction. Normally, natural selection favors low mutation rates, since most mutations are harmful. But in times of stress, extra mutations may raise the odds that organisms will hit on a way out of their crisis. To avoid starving, *E. coli* does not need to know that a small mutation to the switch controlling its lactose-digestion genes will hit the jackpot. It just has to change enough DNA until it changes the right one.

Hypermutation has an obvious risk: along with a beneficial mutation, it can also cause many harmful ones. Susan Rosenberg of Baylor College of Medicine and her colleagues argue that *E. coli* minimizes this risk by spreading it across an entire colony. When *E. coli* produces extra mutations under stress, an individual microbe experiences them only in one narrow region of its DNA. From one microbe to the next, that window of mutation is in a different spot. As a result, the bacteria are not crippled by mutations all across their genome. At the same time, though, new versions of almost every gene in the *E. coli* genome can emerge in a colony. When a few microbes hit on the winning solution, they can start growing quickly.

Hypermutations may be a useful way for *E. coli* to cope with stress, but they may have evolved for very different reasons. Olivier Tenaillon of France's National Institute of Health and Medical Research points out that it takes a lot of energy and material to build hi-fi polymerases. In times of stress, *E. coli* may not be able to afford the luxury of accurate DNA repair. Instead, it turns to the cheaper lo-fi polymerases. While they may do a sloppier job, *E. coli* comes out ahead on balance. Natural selection, Tenaillon proposes, didn't favor higher mutation rates—just cheaper repairs.

Even if the changing mutation rate in bacteria arose as a side effect, it may still be useful. Tenaillon and his colleagues have demonstrated that *E. coli* varies enormously in its mutation rate. Under stress, one microbe may mutate a thousand times faster than another. Hypermutation genes must be responsible for the difference, and they can be passed down from one generation to the next.

In different situations, natural selection may favor some mutation rates over others. Tenaillon and his colleagues have observed the average mutation rate in *E. coli* as it colonizes a mouse. Early on during the colonization, when the bacteria are experiencing a lot of stress, high-mutation microbes are more common. When the bacteria have established stable colonies in the guts of the mice, low-mutating microbes take over. Antibiotics may also drive the rise of high mutators because they can evolve resistance faster than bacteria that mutate more slowly.

Some critics are skeptical of directed mutation, hypermutation, and their intellectual offspring. John Roth of the University of California, Davis, and Dan Andersson of Uppsala University in Sweden argue that Cairns did not discover anything out of the ordinary in his original experiments. The *lac* operons in the bacteria he used were not entirely shut down, Roth and Andersson claim. They could still produce a few proteins, allowing the bacteria to avoid starvation. An ordinary, random mutation might have copied the *lac* operon in a microbe, allowing it to digest more lactose and grow faster. Its descendants might accidentally have made a third copy of the genes, and natural selection might have favored that mutation as well.

Through nothing more than spontaneous mutations and natural selection, Roth and Andersson argue, *E. coli* can expand its collection of lactose-digestion genes. And as the number of copies grows, it becomes more likely that an ordinary mutation will restore one of the operons to good working order. Any microbes that gain a working operon will suddenly multiply far faster than the other bacteria. Mutations may then remove the defective copies, leaving the microbes with a single good version. This process creates the illusion of directed mutations, Roth and Andersson argue, when nothing of the sort has taken place.

The debate, which continues to rage, matters both to the practice of medicine and to our understanding of how life works. If microbes do depend for their survival on an ability to change their mutation rates, then blocking that change could be a way to kill them. Floyd Romesberg has shown that preventing *E. coli* from raising their mutation rate prevents them from evolving resistance. He and his colleagues are now trying to turn that discovery into a medical treatment. They hope that someday people who take antibiotics will also be able to take a drug to stop microbes from increasing their mutations.

Some scientists suspect that animals and plants can also manipulate their mutations to cope with stress. Susan Lindquist of the Whitehead Institute for Biomedical Research in Cambridge, Massachusetts, and her colleagues discovered that fruit flies have a buffer to protect themselves from the harmful effects of mutations. A harmful mutation might cause a protein to fold incorrectly. But the fruit fly's heat-shock proteins can fold it into its proper shape. Over many generations, Lindquist argues, the fruit flies can generate a lot of genetic diversity that could not exist without the help of their heat-shock proteins.

Lindquist discovered that stress unmasks these mutations. Raising the temperature, adding toxic chemicals, or otherwise abusing the flies makes even normal proteins go awry. The heat-shock proteins become so overworked that they abandon many of the mutant proteins to assume their true shapes. These proteins can have drastic effects on the flies, altering their eyes, wings, or other body parts.

Lindquist proposes that heat-shock proteins let the flies build up a supply of mutations that help them survive a crisis without having to suffer their ill effects in less stressful times. An unmasked mutation may prove helpful to the flies, and new mutations can allow it to remain unmasked even after the stress has disappeared. Lindquist and her colleagues have found a similar mutation buffer in plants and fungi, suggesting that it may be a common strategy. The process Lindquist has proposed is different in the details from hypermutation in *E. coli,* but the fundamental benefits seem to be the same: harnessing the creative powers of mutations while minimizing their risks.

Roth and Andersson's gene amplification, on the other hand, may not be limited to a few lactose-starved *E. coli.* Making extra copies of genes may help many organisms adapt to new challenges.

Imagine that a microbe encounters a new kind of food that its ancestors had never tasted. All of the enzymes it uses for feeding have been honed by natural selection for feeding on other molecules. That doesn't necessarily mean the microbe can't eat the new food. Enzymes are actually not all that finely tuned. An enzyme that can slice up one molecule very efficiently may slice up other kinds of molecules, too, albeit more slowly and clumsily. If mutations give a microbe more copies of the gene, it may be able to eat more of the new food.

Ichiro Matsumura, a biologist at Emory University, used *E. coli* to demonstrate just how promiscuous enzymes can be. Matsumura and his colleagues created 104 strains of *E. coli*, each missing a gene that is absolutely essential to the survival of the microbe. They then created thousands of plasmids, each carrying several copies of another *E. coli* gene. After adding these plasmids to the crippled strains, they waited to see if the plasmid genes would be able to pinch hit for the essential gene Matsumura had knocked out. Matsumura found that he could revive 21 out of the 104 strains.

Matsumura's experiment exposed a hidden versality in *E. coli* that may let it adapt to new conditions. Other species may depend on the same potential in their DNA. As mutations make extra copies of those genes, they can do an even better job of feeding on a new food, or detoxifying some poison, or coping with unprecedented heat. In time, one of the copies of the gene may evolve into a far more efficient form. The other genes may then fade away.

Gene amplification may be a creative force, but it can also put us in mortal danger. Like *E. coli*, the cells in our bodies sometimes mutate. On very rare occasions, mutations in our cells put them on the road to becoming cancerous. They no longer obey the controls that keep the growth of normal cells in check. As they continue to divide and mutate, new mutations help them become more aggressive and better able to evade the immune system. Like *E. coli* starving for lactose, these cells face many challenges, and any mutation that helps them overcome these is favored by natural selection. Mutations can create extra copies of genes, which can allow tumor cells to grow faster or escape chemotherapy. Some of these extra genes can evolve new functions of their own that make the tumor even more dangerous.

Sometimes *E. coli* is a little too much like the elephant for the elephant's comfort.

A GIFT OF GENES

World War II, like all wars, provided *E. coli* with a ripe opportunity for slaughter. Its dysentery-causing strains, then known as *Shigella*, stormed across battlefields and invaded cities, killing beyond counting. At the end of the war, *Shigella* retreated from countries that rebuilt their sewers and water supplies. However, in places where water remained dirty—much of Africa, Latin America, and Asia—*Shigella* continued to thrive. The one exception to the rule was Japan. Japan cleaned up its water, and for two years dysentery rates fell. But then, inexplicably, *Shigella* surged back. There were fewer than 20,000 cases in 1948 but more than 110,000 in 1952.

Japanese microbiologists had been very familiar with *Shigella* ever since Kiyoshi Shiga discovered it in 1897. During the postwar outbreak of *Shigella,* they gathered thousands of samples of the bacteria from patients and searched for the cause of its resurgence. Antibiotic resistance, they discovered, was on the rise. At first, microbiologists discovered *Shigella* strains resistant to sulfa drugs. Within a few years, resistance to tetracycline also emerged, then resistance to streptomycin and chloramphenicol.

At first the spread of resistant *Shigella* followed the pattern mapped out in other bacteria, with mutations giving rise to powerful new genes that gave individual microbes a reproductive edge. But then something startling happened. *Shigella* strains emerged that were resistant to *all* the antibiotics. Their transformation was sudden: if doctors gave a victim of *Shigella* a single type of antibiotic, the bacteria often became resistant not only to that drug but to other antibiotics the patient had never taken.

To make sense of this strangeness, Japanese scientists turned to Joshua Lederberg's discovery of sex in *E. coli* a few years earlier. Lederberg had shown that on rare occasion the bacteria could transmit some of their genes to unrelated bacteria. In his experiments, ringlets of DNA—plasmids—moved from one microbe to another, dragging parts of the chromosome with them. Lederberg and other researchers had also discovered that prophages—those quiet viruses—could shuttle genes as well. A roused virus sometimes accidentally copied genes from its host into its own DNA and carried them to other bacteria. Lederberg and other scientists won Nobel Prizes for their discoveries, but for years most scientists considered this "infective heredity" only a convenient laboratory tool. It was not an important part of the natural world. They were wrong, and the dysentery outbreaks in Japan offered the first proof.

Tsutomo Watanabe at Keio University in Tokyo and other Japanese scientists explored the possibility that infective heredity was behind the rise of resistant *Shigella.* They proved that *E. coli* K-12 and *Shigella* could trade resistance genes. Experiments on patients infected with *Shigella* brought similar results. Watanabe concluded that the heavy use of antibiotics had spurred the evolution of resistance genes, either in *Shigella* or in another species of bacteria that lived in the gut. On rare occasion, a resistant microbe passed its genes to another species. These genes, later research would show, were carried on plasmids.

With each new antibiotic that Japanese doctors began to use on their patients, new resistance genes evolved, and their plasmids also spread among the bacteria of Japan. Sometimes a microbe would wind up infected with two plasmids at once, each carrying a gene for resistance to a different drug. The two plasmids swapped genetic material, producing a new ring of DNA carrying two resistance genes instead of one. Natural selection now favored the new plasmids even more, because they allowed bacteria to survive either drug. And over time the plasmids kept picking up other resistance genes. Eventually they made *Shigella* impervious to anything doctors tried to throw at it.

Few scientists outside Japan knew of these discoveries until 1963, when Watanabe wrote a long article in English for the journal *Bacteriological Review*. Western scientists were taken aback. They followed up with experiments of their own and confirmed that Watanabe was onto something big. Genes can shuttle between bacteria by many routes. Plasmids deliver some of them, but viruses deliver them as well. They accidentally incorporate some host genes into their own genome, which the viruses then carry to new hosts that they infect. Sometimes bacteria simply slurp up the DNA that spills out when other microbes die. These resistance genes can shuttle between individuals of the same species, and they sometimes leap from one species to another.

Horizontal gene transfer, as this genetic leaping is now known, works best in places where bacteria are packed in tight quarters. Many genes shuttle between microbes inside our bodies, as well as inside the bodies of chickens and other livestock that are fed antibiotics. Even houseflies that pick up *E. coli* can become a gene market. Horizontal gene transfer allows genes to leapfrog from microbe to microbe across staggering distances. In the jungles of French Guiana, scientists have found antibiotic-resistant *E. coli* in the guts of Wayampi Indians, who have never taken the drugs. In a survey of *E. coli* living in the Great Lakes, another team of scientists discovered resistance genes in 14 percent of them.

Horizontal gene transfer not only spreads resistance genes around but also speeds up their evolution. Once a gene evolves some resistance to antibiotics, it can benefit not just its original host but other bacteria that take it up. And once in its new host, the gene can continue to undergo natural selection and become even more effective. Microbes can assemble arsenals to defend themselves against antibiotics, gathering weapons from the community of bacteria rather than just inheriting them from their ancestors.

Biologists were slow to recognize just how important the *Shigella* outbreak in Japan was. Horizontal gene transfer was helping to create a medical disaster, one that is continuing to unfold. At first biologists did not see much evidence of horizontal gene transfer beyond resistance to antibiotics. In the 1990s, scientists began to compare the entire genome of *E. coli* with that of other bacteria and make a careful search for traded genes. And when they did, our understanding of the history of life changed for good. Horizontal gene transfer, we now know, is no minor trickle of DNA. It is a flood. And it played a big part in making *E. coli* what it is today.

Eight

OPEN SOURCE

A YOUNG SPECIES

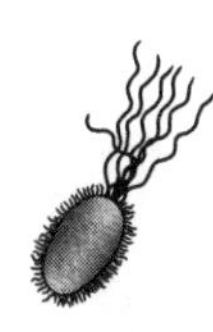

***E. COLI* IS TRAILED BY** thousands of personal historians. They chronicle the birth of sickening new strains in Omaha and Osaka. They trawl streams, lakes, and the guts of kangaroos. They carefully observe the peculiar ways of mutant strains. As the mutants are passed from lab to lab, frozen in stock centers and thawed for new experiments, scientists draw family trees to track their dynasties. Aside from ourselves, we have chronicled no other species so thoroughly.

The written history of *E. coli* is now far too big for any single person to read in a lifetime. But it is both vast and shallow. It begins only in 1885, with Theodor Escherich's first sketches of bunches of rods. Archaeologists can offer a few clues to *E. coli*'s pre-Escherich existence. In 1983, English peat cutters discovered the body of a 2,200-year-old man preserved in a bog near Manchester. The man had been ritually killed: someone had clubbed him on the head, slit his throat, wrapped a cord tightly around his neck, and then pushed him into the bog. The acidic waters preserved his corpse and even its contents. In his stomach, scientists found barley and mistletoe. And in his intestines they found the DNA of *E. coli.*

There's no reason to think that the bog man was the first human ever to carry *E. coli.* There is every reason to think that its history reaches much farther back. Bacteria have an ancient fossil record. Individual microbes left their marks on rocks as least 3.7 billion years ago. Ocean reefs built by bacteria 3 billion years ago still stretch for miles across Africa and Canada. *E. coli* does not do such a good job of forming fossils, because of its tenuous existence. But what *E. coli* lacks in fossils it more than makes up for in the historical record that it carries in its DNA. That genetic record rolls out before us like a carpet, back across millions of years to the origins of *E. coli* as a species, back farther to a time before life dwelled on dry land, back to the origins of cells, to the earliest days of life itself.

To read this record, it's necessary to become a genealogist of bacteria. When a mutation arises in an *E. coli,* it will be passed down to its offspring. That mutation can sometimes serve as a genetic marker, revealing to scientists a group of bacteria that are closely related to one another. It's these genetic markers that public health workers use when an outbreak of nasty *E. coli* occurs, in order to trace the pathogens to their source. Other scientists use these markers to draw branches on *E. coli*'s family tree. They have a long way to go before they finish drawing it, but they've already filled in enough branches to learn some profound things about the bacteria.

All living strains of *E. coli* descend from the first members of the species. Scientists have a rough idea of when those earliest *E. coli* lived. In 1998, Jeffrey Lawrence of the University of Pittsburgh and Howard Ochman of the University of Arizona estimated when the ancestors of *E. coli* and the ancestors of its close relative *Salmonella enterica* split off from each other. Lawrence and Ochman tallied the differences in the species' DNA. When two species branch off from a common ancestor, they acquire mutations at a roughly regular rate. Lawrence and Ochman estimated their common ancestor lived about 140 million years ago. In 2006, Ochman and several other colleagues tackled *E. coli*'s origins from another direction: they surveyed *E. coli* strains and estimated when their common ancestor lived. They concluded that the species was already well established 10 million to 30 million years ago. *E. coli* is much older than the English bog man, in other words, but it is not a living fossil. It is about as primitive as a primate.

E. coli's ancestors split from those of *Salmonella* at a time when dinosaurs dominated the land. Pterosaurs flew overhead, along with birds that still had teeth in their beaks and claws on their wings. The typical mammal at the time was a squirrel-like creature. Around 65 million years ago this picture began to change dramatically. Pterosaurs and the big dinosaurs became extinct, probably in part thanks to an asteroid that crashed into the Gulf of Mexico. After the crash, mammals diversified into flying bats, enormous elephant-like browsers, cat- and doglike carnivores, seed-gnawing rodents, tree-scampering primates. Birds took on their modern forms as well. Mammals and birds share more than survival, however.

Their ancestors independently evolved the ability to control their body temperature. Their guts became a desirable habitat for bacteria, including the ancestors of *E. coli.* Warm-blooded animals need to eat a lot of food to fuel their metabolism, and that rich diet can support a menagerie of microbes. The constant warmth of their guts allows the enzymes of microbes to work quickly and efficiently. It may be no coincidence that the rise of *E. coli* coincides with the rise of its current hosts.

The early *E. coli* produced the vast diversity of lineages that live inside us today, some harmless, some even beneficial, some that ravage the brain or ruin the kidneys, and some that are adapted to life outside warm bodies altogether. Life has often exploded into this sort of diversity when it has gotten the opportunity. But *E. coli*'s explosion is different: scientists can dissect it gene by gene.

VENN GENOMES

Two strains, K-12 and O157:H7, are enough to provide a sense of how diverse *E. coli* is as a species. K-12 is so harmless that scientists make no efforts to protect themselves from it; instead, they have to protect it from fungi and bacteria. If K-12 is a lapdog, O157:H7 is a wolf. It injects molecules into our cells, disrupts our intestines, makes us bleed, loads us with toxins, shuts down our organs, and sometimes kills us. Each microbe relies on a network of genes and proteins to thrive in its particular ecological niche, and those networks are very different from one another. As different as they are, though, K-12 and O157:H7 have a common ancestor, which scientists estimate lived 4.5 million years ago—at a time when our ancestors were upright-walking apes.

In 2001, scientists got their first good look at how a single microbe could give rise to two such different organisms. It was in that year that two teams of scientists—one Japanese, the other American—independently published the complete genome of O157:H7. Scientists could then compare it, gene for gene, with the genome of K-12, which had been published four years earlier. No one could quite predict what would be found.

In the 1970s, scientists had begun comparing small fragments of DNA from different strains of *E. coli.* The fragments were nearly identical from strain to strain, both in their genetic sequence and in their position on the chromosome. Scientists could even find the corresponding fragments in *E. coli*'s relative *Salmonella enterica.* Many scientists assumed that *E. coli*'s entire genome would follow this pattern. They thought *E. coli*'s evolutionary history was tidy. An ancestral microbe had given rise to many lineages, some of which evolved into today's strains. Mutations cropped up in each lineage, a few of which were favored by natural selection, driving their cousins to extinction. Horizontal gene transfer might have imported a few genes from other species, but many scientists assumed that had been a rare event, Lederberg's and Watanabe's work notwithstanding.

But when scientists were finally able to compare the genomes of K-12 and O157:H7, that's not what they found. Vast amounts of DNA in each strain had no obvious counterpart in the other. *E. coli* O157:H7 has 5.5 million base pairs of DNA, while K-12 has only 4.6 million. About 1.34 million base pairs in O157:H7 cannot be found in K-12, and more than half a million base pairs in K-12 have no counterpart in O157:H7. A map of the genes in each genome offered a similar picture. K-12 has 4,405 genes, 528 with no counterpart in O157:H7. Some 1,387 genes in O157:H7 cannot be found in K-12.

Each genome is like a circle in a Venn diagram. The overlap between K-12 and O157:H7 represents a core of shared genes, inherited from a common ancestor. After the two lineages diverged, they acquired new genes from other microbes—not just genes for resistance to antibiotics, but hundreds of other genes that came to make up a quarter of their genomes.

A year after the publication of O157:H7's genome, scientists published the genome of a third strain. Known as CFT073, it lives harmlessly in the intestines, but if it gets into the bladder it can cause painful infections. The scientists discovered that its genome formed a third overlapping circle on the *E. coli* Venn diagram. CFT073 shares some genes with K-12 that it doesn't share with O157:H7. And it shares some genes with O157:H7 that it doesn't share with K-12. But scientists could not find any counterpart for 1,623 of its genes in the other two strains. At the center of the new Venn diagram was the new core of *E. coli* genes. Of all the *E. coli* genes scientists had now identified, only 40 percent could be found in all three strains. The core was shrinking.

As I write this, scientists have sequenced more than thirty *E. coli* genomes; a vast number of other strains are left to examine. With every new strain, scientists continue to discover dozens, even hundreds of genes found in no other *E. coli* strain. Each strain also carries hundreds of genes that it shares with some other strains. The list of genes shared by every *E. coli* is getting shorter, while the list of genes found in at least one strain is getting longer. Scientists call this total set of *E. coli* genes the pangenome. It's up to 11,000 genes now, and at the current rate it will probably become larger than the 18,000 or so genes in the human genome.

The discovery of the *E. coli* pangenome called for a radical rethinking of how the microbe evolved. *Tidy* is precisely the wrong word to describe the history of *E. coli* over the past 30 million years. From the earliest days of its existence, a steady surge of new DNA has entered its genomes. Some of those genes moved from one strain of *E. coli* to another, while some of them came from other species.

Foreign DNA has taken several routes into *E. coli*'s genome. Plasmids, those tiny ringlets of DNA, brought some. Viruses that infect *E. coli* brought more. In some cases, viruses have brought only one or two genes. In other cases, they have brought dozens. These gene cassettes, as they're sometimes called, are not random collections of DNA. They often contain all the instructions necessary to build a complex structure, such as a syringe for injecting toxins. Once these genes become part of the genome of a strain of *E. coli*, the microbes pass them down to their descendants. Ordinary natural selection can fine-tune the genes for the microbe's particular way of life. Sometimes the genes slip away to a new host.

Viruses are quickly losing their reputation as insignificant parasites. They are the most abundant form of life on Earth, with a population now estimated at 10^{30}—a billion billion trillion. Most of the diversity of life's genetic information may reside in their genomes. Within the human gut alone there are about a thousand species of viruses. As viruses pick up host genes and insert them in other hosts, they create an evolutionary matrix through which DNA can shuttle from species to species. According to one estimate, viruses in the ocean transfer genes to new hosts 2 quadrillion times every second.

It's a bizarre coincidence that just as scientists were discovering the evolutionary importance of viruses, computer engineers were creating a good metaphor for their effect. In the late 1990s, a group of American engineers became frustrated by the slow pace of software development. Corporations would develop new programs but make it impossible for anyone on the outside to look at the code. Improvements could come only from within—and they came slowly, if at all. In 1998, these breakaway engineers issued a manifesto for a different way of developing programs, which they called open-source software. They began to write programs with fully accessible code. Other programmers could tinker with the program, or merge parts of different programs to create new ones. The open-source software movement predicted that this uncontrolled code swapping would make better programs faster. Studies have also shown that software can be debugged faster if it is open source than if it is private. Open-source software has now gone from manifesto to reality. Even big corporations such as Microsoft are beginning to open up some of their programs to the world's inspection.

In 2005, Anne O. Summers, a microbiologist at the University of Georgia, and her colleagues coined a new term for evolution driven by horizontal gene transfer: open-source evolution. Vertical gene transfer and natural selection act like an in-house team of software developers, hiding the details of their innovations from the community. Horizontal gene transfer allows *E. coli* to grab chunks of software and test them in its own operating system. In some cases, the combination is a disaster. Its software crashes, and it dies. But in other cases, the fine-tuning of natural selection allows the combination to work well. The improved patch may later end up in the genome of another organism, where it can be improved even more. If *E. coli* is any guide, the open-source movement has a bright future.

ASSEMBLING ASSASSINS

Among its many accomplishments, open-source evolution has produced a lot of ways for us to get sick. When Kiyoshi Shiga discovered *Shigella,* he believed it was a distinct species, and so did generations of scientists who followed him. But when scientists began to examine the genes of *Shigella* in the 1990s, they realized it was just a particularly vicious form of *E. coli.* More detailed comparisons revealed that *Shigella* is actually many separate strains. Many of them are more closely related to harmless strains of *E. coli* than they are to other strains of *Shigella.* In other words, *Shigella* is not a species. It is not even a single strain. It is more a state of being, one that has been achieved by several lineages of *E. coli.*

Shigella strains typically evolved from less sophisticated parasites. Their ancestors sat on top of the cells of the intestinal wall, injecting molecules into host cells to make them pump out fluids. (Many strains of *E. coli* still make this sort of living today.) *Shigella*'s ancestors acquired new genes that allowed them to invade and move inside cells, to escape the immune system and manipulate it. These innovations did not happen in a single lineage of *E. coli.* They evolved many times over.

Just as important as the genes *Shigella* gained were the ones it lost. Flagella are wonderful for swimming in the gut, but they are useless inside the crammed interior of a host cell. No *Shigella* strain can make flagella, although they all still carry disabled copies of the flagella-building genes. *Shigella* also has disabled copies of genes for eating lactose and other sugars that it no longer feeds on. And it has abandoned an enzyme called cadaverine, which other strains of *E. coli* make to protect themselves from acid. (Other bacteria produce this foul-smelling substance as they feed on cadavers; hence the name.) For *Shigella,* cadaverine is a burden because it slows down the migration of immune cells across the wall of the intestines. *Shigella* depends on immune cells to open up passageways that it can use to get into the intestinal tissue and invade cells. As a result, one of the genes essential for making cadaverine has been disabled in every strain of *Shigella.*

Other strains of *E. coli* have evolved into different sorts of pathogens, and their genomes still record that transformation. Horizontal gene transfer, lost genes, and natural selection all were at play in their histories as well. Scientists who study *E. coli* O157:H7, the strain that can be carried in spinach or hamburgers, have done a particularly good job of reconstructing its evolution step by step. Its ancestors started out as far gentler pathogens, but about 55,000 years ago, they began to be infected with a series of viruses, each installing a new weapon in its arsenal. The devastating toxin that makes *E. coli* O157:H7 so dangerous, for example, is encoded on a gene that lies nestled among the genes of a virus. The virus is such a recent arrival in the *E. coli* O157:H7 genome that it still makes new viruses that can escape the microbe.

Scientists who study *E. coli* O157:H7 face a strange paradox, however. Other disease-causing strains of *E. coli,* such as *Shigella,* are highly adapted to living in humans and are rarely found in other species. But *E. coli* O157:H7 is just the opposite. It rarely turns up in humans (for which we can be grateful), but it lives in many cows and other farm animals. In us it can be deadly, but in them it causes no harm at all. It has adapted to them, in other words, as a benign passenger. The fact that O157:H7's toxins make us deathly ill is just an evolutionary accident, because we are not their normal host.

If *E. coli* O157:H7 doesn't make toxins to exploit us, then, why do they carry the toxin genes around? Some researchers have suggested that the bacteria make the toxin to help their animal hosts. At the University of Idaho, scientists have found that sheep infected with *E. coli* O157:H7 do a better job of withstanding a cancer-causing virus than sheep without that strain. They speculate that *E. coli* O157:H7's toxins stimulate the ovine immune system, or perhaps even trigger cells infected with the cancer-causing virus to commit suicide before they can form tumors. But it's also possible that the toxins are a defense for the bacteria themselves. When protozoans attack *E. coli* colonies, the ones that make the toxin can fend off the predators.

While *E. coli* O157:H7 may not have evolved to adapt to our bodies, we have still played a part in its rise. Studies on its genome show that it is a very young lineage; all of its most common forms are less than a thousand years old. Scientists suspect that by domesticating animals, humans created the conditions in which *E. coli* O157:H7 could thrive. Its hosts now spent much of their time penned together on farms, where the *E. coli* O157:H7 that they shed with their manure had a much better chance of infecting a new host than if the cows were off in the wild. *E. coli* O157:H7 exploded with the growth of cattle herding in recent centuries, first with the arrival of cows in the New World and more recently as cows have been packed together into feedlots. The bacteria haven't just become more common thanks to us; they may also have been evolving faster, because viruses have been able to move from microbe to microbe, producing new strains of *E. coli* O157:H7.

While some harmless strains of *E. coli* have evolved into deadly parasites, evolution has flowed the other way as well. Some of the most benign strains of *E. coli* descend from pathogens. One strain of *E. coli,* known as A0 34/86, shields its hosts from invasions of diarrhea-causing bacteria. Doctors sometimes administer it to premature babies to protect their underdeveloped intestines from attack. In 2005, scientists published the genome of A0 34/86. They found genes for cell-killing factors, bloodletting proteins, and other weapons used by O157:H7 and other lethal strains. A0 34/86 uses these dark powers for our good, by aggressively establishing colonies in the guts of babies, thereby preventing disease-causing strains from finding a place to settle. We may try to draw sharp lines through nature's diversity, to split *E. coli* up between its killers and its protectors. But evolution does not deal in sharp lines. It blurs.

ONE LIFE, MANY MASTERS

Another blurred line is the one that divides *E. coli* from its viruses. It may seem sharp if you are looking at *E. coli* ripped open by viruses streaming out to infect a new host. These seem like two different beasts. But *E. coli* has many different kinds of relationships with its viruses. Prophages can, at least for a time, seamlessly blend themselves into their host genomes. They do not necessarily surrender their sense of identity, though. They can sense when their host begins to suffer and at that point they turn back into familiar, host-killing viruses again. And then there are the viruses that lug around bundles of genes that can be very helpful to a microbe but offer no immediate benefit to themselves. When they slip into the genome of *E. coli,* it becomes much harder to say where the virus leaves off and the host begins. The viruses may then become trapped for good inside *E. coli*'s genome, thanks to mutations that destroy their ability to make new copies. Over time, new mutations may chop out much of the virus's original DNA, leaving behind only those genes that are useful to the host. They are viral genes in name and origin only.

To make sense of this confusing relationship between *E. coli* and its viruses, it helps to set aside the "us versus them" view of life and to think of life as a braiding stream of genes. The genes carried by a virus at any moment form a coalition of evolutionary partners that have more success working together than any one of them could have on its own. Some coalitions thrive simply by invading a host and using it to replicate more copies of themselves. But in other cases, the interests of the virus's genes and *E. coli* align. They may enjoy more reproductive success if they spare their host rather than kill it. Some viruses end up as itinerant Samaritans, bringing with them many genes that benefit their host—and, by extension, themselves. They are constantly trying out new combinations of genes as they travel, and the combinations that bring the most success to their hosts are the ones that survive.

These relationships can get complicated, as most relationships do. A virus can be simultaneously benign and malignant toward its *E. coli* host. *E. coli* O157:H7, for example, carries the genes of a virus that include the toxin gene. It's possible that the bacteria benefit from making the toxin, perhaps because it keeps predators at bay; but for the individual microbes that actually produce it, the experience is not so pleasant. The virus forces the microbe to make both toxin molecules and new copies of itself until it bursts.

The decision to make the toxin lies with the virus, not with *E. coli.* It produces the toxin in times of stress—which is one reason why doctors generally don't prescribe antibiotics for an infection of *E. coli* O157:H7. The drugs trigger the viruses to escape their hosts, turning what might have been an unpleasant bout of bloody diarrhea into a potentially lethal case of organ failure. The virus's habit of killing its *E. coli* O157:H7 host is almost enough to inspire pity for the microbe. It is as much a victim of the virus as we are. Even after the viruses have killed their original host, they continue to make things worse for *E. coli.* They infect the harmless strains of *E. coli* in our gut, transforming them into factories that turn out more viruses and more toxins. These hapless bystanders boost *E. coli* O157:H7's production of toxins a thousandfold.

Other viruses use a different strategy to survive, but one that's no less cruel to *E. coli.* Rather than destroying their host when times are bad, they hold it hostage. One of these viruses, known as P1, carries a gene that makes a protein called a restriction enzyme. Restriction enzymes are able to grab DNA at specific sites and slice it apart. Yet P1 normally does not kill *E. coli.* That's because the virus also makes a second protein that protects the microbe from the restriction enzymes. Known as a modification enzyme, it builds shields around *E. coli*'s DNA at exactly the sites where the restriction enzyme can grab it.

Why should P1 bother building both a poison and its antidote? Like many viruses, P1 lives on a plasmid. Each time *E. coli* divides, it usually makes new copies of both its own DNA and the P1 plasmid. Sometimes *E. coli* makes a mistake, however, and all the plasmids end up in one offspring with none in the other. Those plasmid-free bacteria might be able to outcompete the ones that still carry the P1 virus, because they don't have to use extra energy to make virus proteins and copy their DNA. So the P1 virus kills them—even though it's not actually in the bacteria. The deadly beauty of restriction and modification enzymes is that restriction enzymes are durable, whereas modification enzymes are short-lived. If *E. coli* loses the P1 virus, it quickly loses its shields and cannot make new ones. Eventually its DNA becomes vulnerable, and the restriction enzymes move in for the kill. Once *E. coli* is infected with P1, in other words, it can't live without the virus.

Genes for restriction and modification enzymes aren't unique to P1. *E. coli* carries many of them on its chromosome. Ichizo Kobayashi, a geneticist at the University of Tokyo, has argued that they also got their start as selfish genes holding their host hostage. He points out, too, that restriction and modification enzymes could have allowed viruses to battle other viruses trying to take over their host. A new virus invading *E. coli* does not have the shields made by the resident virus, leaving it open to attack by restriction enzymes. While restriction and modification enzymes may have gotten their start as ways to let a parasite thrive, some of them appear to have been harnessed by their *E. coli* hosts. By killing incoming viruses, they have become a primitive sort of immune system for the bacteria.

Genes come into similar conflict in all species. Many insects are infected with a microbe called *Wolbachia,* for example, that can only live inside their cells. It relies for survival almost entirely on being passed down from one generation to the next. This strategy has one major shortcoming: *Wolbachia* cannot infect sperm, and so males are a dead end for its posterity. In other words, the success of *Wolbachia*'s genes and those of its male hosts are in conflict.

Wolbachia has evolved many ways to win this struggle. In some species of wasps, for example, *Wolbachia* manipulates infected females so that they give birth only to females, and it alters their offspring so that they have no need to mate with males to reproduce. In other species, *Wolbachia* kills an infected mother's male eggs. The bacteria in the male eggs die as well, but the strategy ensures the overall success of *Wolbachia* genes: the *Wolbachia*-infected female eggs survive, and when they hatch the female larvae don't face competition for food from their brothers. In fact, their brothers become their food. *Wolbachia,* in other words, has hit on some of the same strategies that viruses use to thrive in *E. coli.*

These murky struggles between parasite and host, these blurrings of species, may seem profoundly alien. Yet we are not above the shaping forces of viruses. Most viruses simply invade our cells, which produce new viruses that move on to the next host. But some viruses insert their genetic material in a cell's genome. If they manage to infect a sperm or an egg, these viruses will be passed down from one generation of humans to the next. Over many generations, mutations cause the viruses to lose their ability to escape their host cells. Many lose most of their genes. What remains are instructions for making copies of their DNA and pasting that DNA back on their host's genome. These genomic parasites now make up about 8 percent of the human genome. Recent research suggests that some of them have been harnessed by their hosts. A number of essential human genes, which help build things as different as antibodies and placentas, evolved from virus genes. Without our resident viruses we would not be able to survive. Once again, what is true for *E. coli* is true for the elephant: Where do our own viruses stop, and where do we begin?

Nine

PALIMPSEST

BURIED MESSAGES

WHEN SCIENTISTS PUBLISHED THE FIRST genome of *E. coli* in 1997, they titled it "The Complete Genome Sequence of *E. coli* K-12." Strictly speaking, the title was a piece of false advertising. Nowhere in the paper can you find the raw sequence of 4,639,221 bases. The omission was simply a matter of space: *E. coli* K-12's genome would fill about a thousand journal pages. Those who crave a direct confrontation with its genetic code must visit the Internet.

One of the sites that houses its code is the Encyclopedia of *Escherichia coli* K-12 Genes and Metabolism, EcoCyc for short. EcoCyc displays the K-12 genome as a horizontal line stretching across the screen, scored with a hash mark every 50,000 bases. If you click on the mark labeled "1,000,000," you will zoom in on the 20,000 bases that straddle that point in the genome. Bars run above the line to show the location of individual genes. Click on the bar for the gene pyrD and you can read its sequence. If you seek something more meaningful, you can also read about pyrD's function (creating some of the building blocks of RNA). On EcoCyc you can learn about the network of genes that controls when pyrD switches on and off.

If you browse EcoCyc for very long, you may fall under a peculiar spell. You may begin to imagine its genome as an instruction manual for an exquisite piece of nanotechnology crafted by some alien civilization. Its genome holds all the information required to assemble and run a sophisticated machine that can break down sugar like a miniature chemical factory, swim with proton-driven motors, and rewire its networks to withstand stomach acids and cold Minnesota winters.

Let that delusion pass.

If you look long enough at *E. coli*'s genome, you will come across hundreds of pseudogenes, instructions with catastrophic typographical errors. You will encounter the genes of viruses that respond to stress by making new viruses and killing their host. Other instructions are mysteriously clumsy, redundant, and roundabout. Still others are cases of outright plagiarism.

Where the metaphor of an instruction manual collapses, other metaphors can take its place. My favorite is an old battered book that sits today in a museum in Baltimore. It was created in Constantinople in the tenth century. A Byzantine scribe copied the original Greek text of two treatises by the ancient mathematician Archimedes onto pages of sheepskin. In 1229, a priest named Johannes Myronas dismantled the book. He washed the old Greek text from the pages with juice or milk, removed the wooden boards, and cut the binding strings on the spine. Myronas then used the sheepskin to write a Christian prayer book. This sort of recycled book is known as a palimpsest.

Despite its new incarnation, the Archimedes palimpsest carried traces of the original text. The prayer book was passed from church to church, scorched in a fire, splashed with candle wax, freshened up with new illuminations, and colonized by purple fungus. In 1907, a Danish scholar named Johan Heiburg discovered that the battered prayer book was in fact the only surviving copy of Archimedes' treatises in their original Greek. But with only a magnifying glass to help him, Heilburg could make out very little of the ancient text. A century later conservationists are making more progress. They are illuminating Archimedes' works with beams of X-rays that light up atoms of iron in the original ink, resurrecting a glowing text of Greek. The palimpsest reveals new depths to the genius of Archimedes, who turns out to have been contemplating calculus and infinity and other concepts that would not be rediscovered for centuries.

E. coli's genome is not so much a manual as a living palimpsest. *E. coli* K-12, O157:H7, and all the other strains evolved from a common ancestor that lived dozens of millions of years ago. And that common ancestor itself descended from still older microbes, stretching back over billions of years. The genetic history of *E. coli* is masked by mutations, duplications, deletions, and insertions. Yet traces of those older layers of text survive in *E. coli*'s genome, like vestiges of Archimedes.

Until recently, scientists had only crude tools for reading those hidden layers. They struggled like Heiberg with his magnifying glass. They are now getting a much better look at the palimpsest. Like Archimedes' ancient treatise, they're finding, *E. coli*'s genome is a book of wisdom. It offers hints about how life has evolved over billions of years—how complex networks of genes emerge, how evolution can act like an engineer without an engineer's brain. Nested within *E. coli*'s genome are clues to the earliest stages of life on Earth, including the world before DNA. Those clues may someday help guide scientists to the origins of life itself.

THE TREE OF LIFE

To read *E. coli*'s palimpsest, scientists have had to figure out which parts of its genome are new and which are old. The answer can be found in the genealogy of germs. A family tree of the living strains of *E. coli* indicates that they all descend from a common ancestor that lived some 10 million to 30 million years ago. Even farther back, *E. coli* shares an ancestor with other species. Reach back far enough, and you ultimately encounter the ancestor *E. coli* shares with *all* other living things, ourselves included.

Reconstructing the tree of life—one that includes *E. coli* and humans and everything else that lives on Earth—has been one of modern biology's great quests. In 1837, Charles Darwin drew his first version of the tree of life. On a page in his private notebook he sketched a few joined branches, each with a letter at its tip representing a species. Across the top of the page he wrote, "I think."

The fact that species have common ancestors explains why they share many traits. As different as bats and humans may seem, we are both hairy, warm-blooded, five-fingered mammals. Darwin himself did not try to figure out exactly how all the species alive were related to one another, but within a few years of the publication of *The Origin of Species,* other naturalists did. The German biologist Ernst Haeckel produced gorgeous illustrations of trees sprouting graceful bark-covered boughs. His trees were accurate in many ways, scientists would later find. But Haeckel marred them with a stupendous anthropocentrism. To Haeckel, the history of life was primarily the history of our own species. His tree looked like a plastic Christmas tree, with branches sticking out awkwardly from a central shaft. He labeled the base of the tree *Moneran,* the name he used for bacteria and other single-celled organisms. Farther up the tree were branches representing species more and more like ourselves—sponges, lampreys, mice. And atop the tree sat *Menschen.*

This view of life has been a hard one to shake. It probably had something to do with the decision to split life into prokaryotes and eukaryotes, the supposedly primordial bacteria and the "advanced" species like ourselves that evolved from them. It's a deeply flawed view. The evolution of life was not a simple climb from low to high. *E. coli* is a species admirably adapted to warm-blooded creatures that did not emerge for billions of years after life began. It is as modern as we are.

It took a long time for a more accurate picture of the tree of life to take hold. One major obstacle was the lack of information scientists could use to determine how *E. coli* is related to other bacteria, or how bacteria are related to us. To compare ourselves to a bat, we can simply use our eyes to study fur, fingers, and other parts of our shared anatomy. Under a microscope, however, many bacteria look like nondescript balls or rods. Microbiologists sometimes classified species of bacteria based on little more than their ability to eat a certain sugar, or the way they turned purple when they were stained with a dye. It was not until the dawn of molecular biology that scientists finally got the tools required to begin drawing the tree of life. Experiments on *E. coli* helped them to recognize that all living things share the same genetic code, and the same way of passing on genetic information to their descendants. They share these things because they had a common ancestry.

In the 1970s, Carl Woese, a biologist at the University of Illinois, Urbana-Champaign, discovered a way to use those shared molecules to draw a tree of life. Woese and his colleagues teased apart ribosomes, the factories for making proteins, and studied one piece of RNA, known as 16S rRNA. Woese did his work years before scientists could easily read the sequence of RNA or DNA. So he and his colleagues did the next best thing: they sliced up *Escherichia coli*'s 16S rRNA with the help of a virus enzyme. They then cut up the 16S rRNA of other microbes and gauged how similar their fragments were to those of *E. coli.* They discovered many regions that were identical, base for base, no matter which species they compared. These regions had not changed over billions of years. The regions that had diverged revealed which species were more closely related than others.

Rough and preliminary as the results were, they upended decades of consensus. The standard classifications of many groups of bacteria turned out to be wrong. Most startling of all, Woese and his colleagues found that a number of bacteria were closer to eukaryotes than to other bacteria. They were not bacteria at all. Woese and his colleagues declared that life formed not two major groups of species but three. They dubbed the third domain of life archaea. "We are for the first time beginning to see the overall phylogenetic structure of the living world," Woese and his colleagues declared.

Over the next thirty years, scientists built on Woese's work, drawing a more detailed picture of the tree of life. They studied ribosomal RNA in more species. They found other genes that also made for good comparisons. They used new statistical methods that gave them more confidence in their results. They found many more species of archaea, confirming it as a genuine branch of life. Archaea may look superficially like bacteria, but they have some distinctive traits, such as unique molecules that make up their cell walls.

To measure the diversity of life, Woese and his colleagues counted up the mutations to ribosomal RNA that had accumulated in each branch of life. The more mutations, the longer the branch. The new tree was a far cry from Haeckel's. The animal kingdom became a small tuft of branches nestled in the eukaryotes. Two bacteria that might look identical under a microscope were often separated by a bigger evolutionary gulf than the one that separates us from starfish or sponges. One look at the tree made it clear that the evolutionary history of any individual species of bacterium—*E. coli,* for example—is a complicated tale.

TREE VERSUS WEB

In the 1980s, some experts on the tree of life became worried. It was slowly becoming clear that horizontal gene transfer was not just a peculiarity of *E. coli*'s laboratory sex or the modern era of antibiotics. Genes had moved from species to species long before humans had begun to tinker with life. If genes moved too often, some scientists feared, they might make it impossible to reconstruct the tree's branches.

To reconstruct the tree of life, scientists compare DNA from different species and come up with the most likely pattern of branches that could have produced the differences. A genetic marker shared by two species might reveal that they had a close common ancestry, one not shared by species that lack the marker. But those markers make sense only if life passes down all its genes from one generation to the next. If a gene slips from one species to another, it can create an illusion of kinship that's not actually there.

At first, most scientists dismissed this sort of fretting. Over the course of billions of years, horizontal gene transfers were inconsequential. To find the true tree of life, scientists assumed they just had to avoid those rare swapped genes.

In later years it became possible to get a better sense of how much horizontal gene transfer has occurred by comparing genomes. The genomes of humans and other animals didn't show much evidence of recently transferred genes. That's not too surprising when you consider how we reproduce. Only a few cells in an animal—eggs and sperm cells—have a chance to become a new organism. And these cells have very little contact with other species that might bequeath DNA to them. (The chief exceptions to this rule are the thousands of viruses that have inserted themselves in our genomes.) But in this respect, animals were oddities. Bacteria, archaea, and single-celled eukaryotes turned out to have traded genes with surprising promiscuity. And those traded genes, some scientists argued, posed a serious threat to the dream of drawing the full, true tree of life.

W. Ford Doolittle, a biologist at Dalhousie University in Halifax, Nova Scotia, illustrated the seriousness of the threat in an article in *Scientific American* in 2000. The article includes a picture of two trees. The first shows the tree of life as revealed by ribosomal RNA, with bacteria, archaea, and eukaryotes branching off in an orderly fashion from a common ancestor. The second shows what the history of life might really look like: a tree emerging from a mangrovelike network of roots, with branches fused into a tangle of shoots. Parts of it look less like a tree than a web.

As with most scientific debates in biology, the tree-versus-web debate is not an all-or-nothing battle. The web champions, such as Doolittle, don't deny that organisms are related to one another by common descent. They just think that searching for one true tree of life by comparing genes is a futile quest. The tree champions do not deny that horizontal gene transfer happens or that it is biologically important. They simply argue that the right genes can reveal the true relationships among all living things on Earth.

As scientists have begun to compare the entire genomes of many species for the first time, a number of them have decided that the tree of life still stands. Howard Ochman came to this conclusion on the basis of a survey he and his colleagues made of *E. coli* and a dozen other species of bacteria. The scientists found a number of genes that showed signs of having moved by horizontal transfer. But most of those genes had moved relatively recently—only after each species in their study had branched off from the others.

Horizontal gene transfer is common, the scientists found, but the genes usually don't survive very long in their hosts. Many of them become disabled by mutations, turning into pseudogenes. Eventually, other mutations slice the genes out of their genomes completely, and the bacteria suffers no ill effects from the loss. A few genes ferried into the ancestors of *E. coli* and other bacteria did manage to establish themselves and can still be found in many living species today. But in order to avoid oblivion, they seem to have abandoned their wandering ways. Once a virus inserted them into a host genome, they did not leave it again. Ochman and his colleagues concluded that even though genes regularly move between the branches of life, the branches remain distinct.

THE ROAD TO *ESCHERICHIA*

The newest versions of the tree of life look nothing like Haeckel's Christmas tree. Scientists can now compare thousands of species at once, and the only way to draw all of their branches is to arrange them like the spokes on a wheel. At the center of the wheel is the last common ancestor of all life on Earth today. From the center you can move outward, steering from branch to branch to follow the evolution of a particular lineage. To get to our own species, you first travel up to the common ancestor of archaea and eukaryotes. From there you bear right onto the eukaryote branch. Our ancestors remained single-celled protozoans until about 700 million years ago. They parted ways with the branches that would give rise to multicellular plants and fungi. Eventually the path takes you to the animal kingdom. Bear right again and you follow our ancestors as they become vertebrates. The ancestors of other vertebrates branch off along the way: zebrafish, chickens, mice, chimpanzees. Finally the line ends with *Homo sapiens.*

But enough about you. A different route travels from the common ancestor to *E. coli.* The journey is just as long and no less interesting.

The last common ancestor of all living things was probably much simpler than *E. coli.* While each species today carries some unique genes, it also shares genes found in all other species. These universal genes probably are the legacy of the last common ancestor. A simple search for universal genes brings up a pretty short list, about 200 genes long. The common ancestor probably had a bigger genome, because many genes have been lost over the history of life. Christos Ouzounis and his colleagues at the European Bioinformatics Institute in Cambridge estimate that its full genome contained somewhere between 1,000 and 1,500 genes. Even if Ouzounis is right, however, the last common ancestor of all living things had only a third or a quarter of the genes that a typical strain of *E. coli* has today.

That last common ancestor did not have early Earth all to itself. It shared the planet with an uncountable number of other microbes. Over time the other branches on the tree of life became extinct while our own survived. The world on which these early microbes lived was profoundly different from our own. Four billion years ago, Earth was regularly devastated by gigantic asteroids and miniature planets. Some of the impacts may have boiled off the oceans. As the water slowly fell back to Earth and grew into seas again, life may have found refuge in cracks in the ocean floor. It may be no coincidence that on the tree of life some of the deepest branches belong to heat-loving species that live in undersea hydrothermal vents.

Once Earth became more habitable, the descendants of the common ancestor fanned out. They spread across the seafloor, growing into lush microbial mats and reefs. Continents swelled up, and early organisms moved ashore, forming crusts and varnishes. Along the way they evolved new ways to feed and grow. Some bacteria and archaea consumed carbon dioxide and used iron or other chemicals from deep-sea vents as a source of energy. They built up a supply of organic carbon that other microbes began to feed on.

E. coli may descend from those ancient scroungers. Its ancestors certainly could not have been living inside humans 3 billion years ago, or inside any other animal for that matter. Some of *E. coli*'s closest living relatives (a group collectively known as gamma-proteobacteria) offer some clues to what *E. coli*'s ancestors might have been doing then. Some eat oil that oozes from cracks in the seafloor. Others live on the sides of undersea volcanoes, where they glue themselves to passing bits of proteins. *E. coli* may have acquired its metabolism from such carbon-scrounging ancestors.

E. coli's complex social life—forming biofilms, waging wars with colicins, and so on—may have also had its origins in free-living ancestors in the ocean. Aquatic microbes today have intensely social lives, living mainly in biofilms rather than floating alone as individuals.

About 2.5 billion years ago, the ancestors of *E. coli* were rocked by a planetwide catastrophe: oxygen began to build up in the atmosphere. To us oxygen is essential to life, but on the early Earth it was poison. Initially the planet's atmosphere was a smoggy mix of molecules, including heat-trapping methane produced by bacteria and archaea. Free oxygen was rare, in part because the molecules rapidly reacted with iron and other elements to form new molecules. Life changed the planet's chemistry when some bacteria evolved the ability to capture sunlight. They gave off oxygen as waste, and after 200 million years it began to build up in the atmosphere. Unless an organism can protect itself, oxygen can be lethal. Thanks to its atomic structure, oxygen is eager to attack other molecules, wresting away atoms to bond with. The new oxygen-bearing molecules can roam through a cell, wrecking DNA and other molecules they encounter.

For the first billion and a half years of life, the planet had been mercifully free of the oxygen menace. And then, 2.5 billion years ago, oxygen levels rose tenfold. The oxygen revolution may have driven many species extinct, while others found refuge in places where oxygen levels remained low—deep inside mudflats, for example, or at the bottom of the ocean. But some species, including the ancestors of *E. coli,* adapted. They acquired genes that protected them from oxygen's toxic effects. Once shielded, their metabolism evolved to take advantage of oxygen, using it to get energy out of their food far more efficiently than before. Today *E. coli* can still switch back and forth between its ancient oxygen-free metabolism and its newer network, depending on how much oxygen it senses in its environment.

The other major revolution that *E. coli*'s ancestors experienced was delivered by our own ancestors. Early eukaryotes, biologists suspect, were the predators of the early Earth. They were much like amoebas today, which prowl through soil and water in search of prey they can engulf. Bacteria that could defend themselves against these predators were favored by natural selection. Today bacteria have an impressive range of defenses against amoebas and other eukaryote predators. They can produce toxins that they can inject with microscopic needles into the amoebas. Their mucus-covered biofilms are difficult for predators to penetrate. Even when ingested, bacteria can avoid destruction.

In some cases, bacteria may have turned the tables on their predators. Amoebas today get sick with bacterial infections caused by species that have evolved the ability to infect and thrive inside hosts. Some bacteria are more polite lodgers, providing single-celled protozoans with life-giving biochemistry. Early eukaryotes acquired oxygen-breathing bacteria this way, and those bacteria are still part of our own cells today. Algae acquired photosynthesizing bacteria, and among their descendants are the plants that make the land green. Thanks to these bacterial partners, the continents could begin to support a massive ecosystem, with forests and grasslands and swamps becoming home to animals of all sorts, from insects to mammals.

These animals, the descendants of the predatory eukaryotes that harassed bacteria billions of years earlier, now became a new ecosystem for bacteria to invade. Thousands of species of microbes, including the ancestors of *E. coli*, adapted to the food-rich realm of the animal gut. They brought with them their abilities to break down organic carbon, communicate with one another, and cooperate. They had come a long way from the common ancestor of all living things. But as they took up residence inside humans and other animals, they had in their own way brought some branches of the tree of life together again.

E. COLI GOES TO COURT

The federal courthouse in Harrisburg, Pennsylvania, is a nondescript box of dark glass. Its judges deal mostly in humdrum conflicts over funeral-parlor regulations, liquor-store licenses, airport parking lots. But in 2005 a surge of people—reporters, photographers, and onlookers—hit the courthouse like a rogue wave. One case had drawn them all: *Kitzmiller v. Dover Area School District.* Eleven parents from the small town of Dover had taken their local board of education to court. They charged that the board was introducing religion into science classes. The world's attention was fixed on the case because it represented the first time the courts would consider creationism in its latest incarnation, known as intelligent design. The trial opened on September 26, 2005. Projectors had been brought into the court, and the lawyers and expert witness used them to display images on a large screen. Again and again the same image appeared: the flagellum of *E. coli.*

Over the past twenty-five years *E. coli*'s flagellum has become an icon to creationists, a molecular weapon they try to wield against the evils of Darwin and his followers. For decades they have touted it in lectures and books as a clear-cut example of the handiwork of a divine designer. But it was not until the Dover case that they had the opportunity to present the flagellum to the world.

The strategy failed miserably. At the end of the trial, Judge John E. Jones ruled against the school board, in part because its case for the flagellum's intelligent design was so weak. In fact, flagella are a fine example of how evolution works and a clear demonstration of why creationism fails as science.

Creationism—the belief that life's diversity originated from specific acts of divine creation—first emerged in American history during the early years of the twentieth century. But it was never a single body of ideas. Some creationists argued that the world was a few thousand years old, while others accepted the geologic evidence of its great age. Some claimed evolution must be wrong because it did not accord with the Bible. Others tried to attack the evidence for evolution. They claimed that living species were so distinct from one another that they could not have evolved from a common ancestor. They pointed out the absence of transitional fossils, such as ones that might link whales to land mammals, and claimed that such gaps were proof that intermediate forms could not possibly have existed. When paleontologists discovered fossils of some of those transitional forms—such as whales with legs—the creationists simply retreated to another gap.

While creationists failed in the scientific arena, they had more luck in public high schools. In the 1920s, state legislatures began banning the teaching of evolution, and many of those laws stayed on the books for more than thirty years. It was not until 1968 that the U.S. Supreme Court ruled that banning the teaching of evolution amounted to imposing religion on students. If creationists could not keep evolution out of classrooms, they would try to get creationism in alongside it. They began to claim that creationism is sound science that deserves to be taught. These self-styled "creation scientists" founded organizations with august names, such as the Institute for Creation Research. They began working on a textbook about creation science that they wanted introduced into schools. And they looked around the natural world for things they could claim as scientific evidence of creation.

Biology had changed dramatically since the birth of creationism. Molecular biologists were plunging into the exquisite complexity of cells, discovering clusters of proteins working together like the parts of machines. Creation scientists mined the new research for structures they claimed were the result of creation, not evolution. One of the things they chose was *E. coli*'s flagellum.

In 1981, Richard Bliss, chairman of the Education Department of the Institute for Creation Research, came to West Arkansas Junior College to give a talk about creation science. He told his audience that in the creation model of life, "we would predict that we'd see a fantastic amount of orderliness, and there is, folks. There's orderliness on a macro level and on a micro level. The further we get down into the molecular level the more we see this orderliness jump out and scream out at us." As an example of this order, Bliss showed his audience a picture of *E. coli.*

Bliss described its flagellum, detailing the many proteins that make it up and how they work together to make it spin. "I like to call it a Mazda engine," Bliss said. He hoped that students could be taught the "creation model" of *E. coli*'s flagellum along with the "evolution model" and make up their own minds. "It's just exciting science and exciting education," he said.

This sort of argument swayed some state legislatures to pass laws requiring that creation science be taught alongside evolution. But the Supreme Court struck the laws down in the 1980s because they, in effect, endorsed religion. The Court declared creation science no science at all.

Creationists repackaged their old claims once more. They stripped away all mention of creationism, creation, and a creator. They argued instead that life shows signs of something they called intelligent design. DNA and proteins and molecular machines are simply too complex to have evolved by natural selection, they argued. These molecules were purposefully arranged, and that purpose reveals an intelligent designer at work. Just what or who that designer is they would not say, at least not publicly.

One of the most striking examples of this makeover was the transformation of a textbook originally called *Creation Biology.* A Texas publishing house had started work on the manuscript in the early 1980s, but in the wake of the Supreme Court's rulings its editors began to replace the words *creationism* with *intelligent design, creator* with *intelligent designer,* and *creationist* with *design adherent.* Otherwise, they barely changed the language. In 1989, the textbook was published. Instead of *Creation Biology,* its publishers named it *Of Pandas and People.*

The evidence for creation—including the flagellum—now became the evidence of intelligent design. Richard Lumsden of the Institute for Creation Research waxed rhapsodic about it in a 1994 article published in the journal of the Creation Research Society, a "young-Earth creationism" organization: "In terms of biophysical complexity, the bacterial rotor-flagellum is without precedent in the living world," Lumsden wrote. "To the micromechanicians of industrial research and development operations, it has become an inspirational, albeit formidable challenge to the best efforts of current technology, but one ripe with potential for profitable application. To evolutionists, the system presents an enigma; to creationists, it offers clear and compelling evidence of purposeful intelligent design."

While some proponents of intelligent design continued to call themselves creationists, others noisily rejected the name. They claimed that intelligent design is only the scientific search for evidence of design in nature. And for them *E. coli*'s flagellum was also a favorite example. William Dembski, a philosopher at Southwestern Baptist Theological Seminary, put it on the cover of his book *No Free Lunch.* He presented a calculation of the probability that *E. coli*'s flagellum had come together by chance. The number he came up with was spectacularly tiny, which Dembski took as evidence that it must have been produced by a designer. Biologists and mathematicians alike reject Dembski's argument because it is supremely irrelevant. Mutations may be random—at least insofar as they don't produce only variations an organism actually needs—but natural selection is not a matter of chance.

Dembski and other proponents of intelligent design claimed that the designer might be an alien or a time traveler. But personally they believed the designer to be God. Dembski wrote that intelligent design is essentially the theology of John's Gospel in the Christian scriptures. And all the talk of aliens and time travelers did not scare off conservative religious organizations. Instead, they embraced intelligent design. Focus on the Family, for example, a large American evangelical organization, urged its members to demand that *Of Pandas and People* be used in schools whenever evolution was taught. In 2002, Focus on the Family's magazine ran an article by Mark Hartwig extolling intelligent design. More than twenty years after Bliss's lecture in Arkansas, creationists were still picking out *E. coli* as one of their prime exhibits.

"Darwinists dismiss the reasoning behind the intelligent-design movement, contending that living organisms were produced by the mindless processes of random mutation and natural selection," Hartwig wrote. "But advances in molecular biology are shredding that claim. For example, consider the little outboard motor that bacteria such as *E. coli* use to navigate their environment. This water-cooled contraption, called a flagellum, comes equipped with a reversible engine, drive shaft, U-joint and long whip-like propeller. It hums along at 17,000 rpm." Hartwig pointed out that it took fifty genes to create a working flagellum. If a single gene were disabled by a mutation, the flagellum would be crippled. There were therefore no intermediate steps by which a flagellum could have evolved. "Such systems simply defy Darwinist explanations," Hartwig declared.

Focus on the Family was not the only organization trying to get *Of Pandas and People* into public schools. In 2000, a Christian legal organization called the Thomas More Law Center began sending lawyers to school boards around the country. They urged the boards to adopt the book and promised to defend them if they were sued. "We'll be your shields against such attacks," Robert Muise, one of the lawyers, told the Charleston, West Virginia, Board of Education. (The Thomas More Law Center calls itself "the sword and shield for people of faith.") School boards in Michigan, Minnesota, West Virginia, and other states turned them down.

But in 2004 the Thomas More lawyers got a break when they visited the rural community of Dover, Pennsylvania. The Dover Board of Education decided to promote the teaching of intelligent design. One board member arranged for sixty copies of *Of Pandas and People* to be donated to the school library. The local school board added a new statement to the science curriculum. "Students," it read in part, "will be made aware of gaps/problems in Darwin's Theory and of other theories of evolution including, but not limited to, intelligent design."

The board of education also demanded that teachers read a second statement aloud to all Dover biology classes. They were required to say that evolution was a theory, not a fact (confusing the nature of both facts and theories). "Intelligent Design is an explanation of the origin of life that differs from Darwin's view," the statement continued. "The reference book *Of Pandas and People* is available for students to see if they would like to explore this view in an effort to gain an understanding of what Intelligent Design actually involves. As is true with any theory, students are encouraged to keep an open mind."

Dover's science teachers refused to read the statement. They declared that to do so would violate the oath they took not to give their students false information. The superintendent came to the classrooms to read the statement instead. When curious students asked what sort of designer was behind intelligent design, he told them to ask their parents and walked out.

Two months later, eleven parents filed a lawsuit. Their lawyers argued that the statement violated the First Amendment because it represented the impermissible establishment of religion. And on an autumn day the trial began.

The plaintiffs called parents and teachers to testify how the board of education had pressured teachers not to teach "monkey to man evolution" and promised to bring God back into the classroom. The defense responded by bringing in two biologists as expert witnesses, Scott Minnich of the University of Idaho and Michael Behe of Lehigh University. Like Dembski, Minnich and Behe are fellows at the Discovery Institute in Seattle, the leading organization for the promotion of intelligent design.

Behe has never managed to publish a paper in a peer-reviewed biology journal arguing for intelligent design based on original research. Instead, he has presented his case mainly in op-ed columns, speeches, and books. Behe claims that some biological systems could not have evolved by natural selection because they are what he calls "irreducibly complex." He asserts that something could be called irreducibly complex if it is "a single system composed of several well-matched, interacting parts that contribute to the basic function, wherein the removal of any one of the parts causes the system to effectively cease functioning." It would be impossible, in his view, for natural selection to gradually produce an irreducibly complex system, because it would have to start with something that didn't work. "If a biological system cannot be produced gradually it would have to arise as an integrated unit, in one fell swoop," he concludes.

Behe uses a few examples to illustrate irreducible complexity. The flagellum is one of his favorites. He claims it is obviously too complex to have evolved from a simpler precursor. Faced with the wonder of the flagellum, Behe writes, "Darwin looks forlorn."

At the Dover trial, Behe had a textbook illustration of *E. coli*'s flagellum projected on the courtroom screen, and he proceeded to marvel at it all over again. "We could probably call this the Bacterial Flagellum Trial," a lawyer for the school board said.

Behe inventoried the flagellum's many parts and told Judge Jones that Darwinian evolution could not have produced its irreducible complexity. "When you see a purposeful arrangement of parts, that bespeaks design," he said. The flagellum, Behe explained, was built for a purpose—to propel bacteria—and it was built from many interacting parts, just like the outboard motor of a boat. "This is a machine that looks like something that a human might have designed," he said.

The plaintiffs' witnesses were eager to talk about the flagellum as well, in order to demolish Behe's claims about irreducible complexity. Kenneth Miller, a biologist at Brown University, pointed out that Behe's claims about irreducible complexity could be tested. Behe, Miller reminded the court, had defined an irreducibly complex system as one that would be nonfunctional if it were missing a part. Miller then showed the court a computer animation of the flagellum. He began to dismantle it, removing not just one part but dozens. The filament disappeared. The universal joint vanished. The motor slipped away. All that was left when Miller was done was the needle that injects new parts of the filament into the shaft.

Miller had removed a great deal of an irreducibly complex system. By Behe's definition, what remained should no longer be functional. But it is. The ten proteins that make up the needle are nearly identical in both their sequence and their arrangement to a molecular machine known as the type III secretion system. This is the needle used by *E. coli* O157:H7 and other disease-causing strains to inject toxins into host cells.

"We do break it apart, and lo and behold, we find—actually we find a variety of useful functions, one of which I have just pointed out, which is type III secretion," Miller testified. "What that means, in ordinary scientific terms, is that the argument that Dr. Behe has made is falsified, it's wrong, it's time to go back to the drawing board."

Behe tried to play down Miller's testimony. When Behe said that a system became nonfunctional when it lost a part, he now claimed, he had meant that it lost its *particular* function. By removing part of the flagellum, Behe argued, Miller was left with something that could not propel a microbe. "If you take away those parts, it does not act as a rotary motor," Behe said.

He then claimed that most people would assume Miller was implying that a type III secretion system evolved into a flagellum, something evolutionary biologists were not agreed on. Some had raised the possibility that the flagellum had evolved into a type III secretion system or that both structures evolved from a common ancestor. Yet Miller had not said anything of the sort. He had simply tested Behe's claims, carefully hewing to Behe's own words. And Behe's claims had not held up to the evidence.

Over the course of the trial it became clear that Behe had some strange demands for scientists who would explain how the flagellum—or any other supposedly irreducibly complex systems—evolved. "Not only would I need a step-by-step, mutation-by-mutation analysis," he said, "I would also want to see relevant information such as what is the population size of the organism in which these mutations are occurring, what is the selective value for the mutation, are there any detrimental effects of the mutation, and many other such questions."

For the flagellum, Behe offered evolutionary biologists an idea for an experiment to overturn irreducible complexity. "To falsify such a claim, a scientist could go into the laboratory, place a bacterial species lacking a flagellum under some selective pressure, for mobility, say, grow it for 10,000 generations, and see if a flagellum, or any equally complex system, was produced. If that happened, my claims would be neatly disproven."

Behe was cross-examined by Eric Rothschild, one of the lawyers for the Dover parents. Rothschild pointed out the inconsistencies riddling his testimony. Behe's proposal for evolving a flagellum in the lab revealed an indifference to the scale of evolution. A 10,000-generation experiment might last two years, whereas bacteria have been evolving for well over 3 billion years. In a typical experiment a scientist might study several billion microbes. But the world's population of microbes is inconceivably larger. A microbe's failure to evolve in a laboratory would offer no evidence of intelligent design.

While Behe issued absurd demands to evolutionary biologists, he demanded little of himself. He felt no need to offer his own step-by-step account of how an intelligent designer created the flagellum (or when, or where, or why). Intelligent design, he informed the court, "does not propose a mechanism in the sense of a step-by-step description of how those structures arose." The only feature that Behe needed to find in those structures to call them intelligently designed was the appearance of design. "When we see a purposeful arrangement of parts, we have always found that to be design," he testified. "What else can one go with except on appearances?"

This sort of testimony persuaded Judge Jones that intelligent design was scientifically empty. In December 2005, he ruled that *Of Pandas and People* had no place in the Dover classroom. "The evidence at trial demonstrates that ID [intelligent design] is nothing less than the progeny of creationism," Jones declared in his decision. He chose the flagellum as an illustration of how seamlessly creationism and intelligent design were connected. "Creationists made the same argument that the complexity of the bacterial flagellum supported creationism as Professors Behe and Minnich now make for ID," he wrote.

The Dover trial was a creationist disaster. The Dover School Board members who had brought *Of Pandas and People* into the school were defeated by a slate of opponents to the policy even before the trial was over. Other intelligent design–friendly board of education members have lost their seats in Kansas and Ohio. Judge Jones's decision was so thorough that it will probably set a precedent for any future cases on the teaching of creationism in whatever guise it takes.

Remarkably, though, creationists still love *E. coli.* Access Research Network, another organization that promotes intelligent design, has plastered its flagellum on T-shirts, aprons, beer steins, baseball jerseys, coffee mugs, calendars, greeting cards, calendars, tote bags, and throw pillows. All these creationist items can be purchased on a Web site. The site declares: "The output of this mechanism is used to drive a set of constant torque proton-powered reversible rotary motors which transfer their energy through a microscopic drive train and propel helical flagella from 30,000 to 100,000 rpm. This highly integrated system allows the bacterium to migrate at the rate of approximately ten body lengths per second. Would you please find out who filed the patent on this thing?"

The message you'll actually get on your flagellum apron will be far simpler. Above the picture of the flagellum it reads, "Intelligent Design Theory." And below: "If it looks designed, maybe it is."

THE FLAGELLUM AFTER DOVER

It was a delicious coincidence that the Dover trial, which brought *E. coli*'s flagellum to the world's attention, took place right around the time scientists were starting to get a good look at the flagellum's evolution. They began to trace the history of its genes by finding related genes both in *E. coli* and in other microbes. Together those genealogies are beginning to add up to a history of the flagellum—and an illustration of how life can produce a complex trait.

The most important lesson of this new research is that it's absurd for creationists to talk of *the* flagellum. From species to species there's a huge amount of variation in flagella. Even within a single species different populations of microbes may make different kinds of flagella.

Flagella vary at all levels, from their finest features to their biggest. Take flagellin, the protein that *E. coli* uses to build the tail of its flagella. Scientists have identified forty kinds of flagellin in various strains of *E. coli,* and they expect to find many more as they expand their survey. And from species to species, flagellins vary even more. In 2003, a ship of microbiologists and geneticists trawled microbes in the Sargasso Sea and analyzed their genes. They discovered 300 genes for flagellins.

These patterns make eminent sense in light of evolution. A single ancestral flagellin gave rise to many new flagellins through gene duplication and mutations. As different species adapted to different environments—from feeding inside the human gut to swimming the Sargasso Sea—their flagellins evolved as well. After *E. coli* emerged tens of millions of years ago, its flagellins continued to evolve. The variation in its flagellins was probably driven by the need to evade the immune system of its host, which recognizes intruders by the proteins on their surface, such as flagellin. If a mutation makes the outer surface of flagellin harder for an immune system to recognize, it may be favored by natural selection. And just as you'd expect, the most variation found in flagellins in *E. coli* lies in the parts that face outward. The parts that face inward—and have to lock neatly into the other flagellins—are much more similar to one another. Natural selection does not look kindly on mutations that disturb their tight fit.

Flagella also vary in other ways. *E. coli* drives its motors with protons, but some species use sodium ions. *E. coli* spins its flagella through a fluid. Other species make flagella for slithering across surfaces. Scientists have observed some species of bacteria that can make either kind, depending on what sort of swimming they have to do.

In 2005, Mark Pallen of the University of Birmingham in England and his colleagues discovered a set of genes for building slithering flagella in an unexpected place: the genome of *E. coli. E. coli* cannot actually build these slithering flagella, because the switch that turns on the genes was disabled by a mutation. In some strains, scientists have found all forty-four genes necessary for building all the parts of the slithering flagellum—its hooks, its rings, its filament. In other strains, some of the genes have disappeared entirely. In K-12 only two badly degraded genes remain.

Pallen's discovery makes ample sense if flagella are the product of evolution, and it makes no sense at all if they are the result of intelligent design. A complex feature evolves and is passed down from ancestors to descendants. In some lineages it falls apart. Darwin described many rudimentary organs, from the flesh-covered eyes of a cave fish to the stubby wings of ostriches. If natural selection no longer favored their use, Darwin argued, individuals would be able to survive well enough even if the organs no longer served their original function. *E. coli* carries vestiges as well, like ancient passages hidden in a palimpsest.

E. coli also carries clues to how its flagellum evolved in the first place. As Kenneth Miller pointed out in the Dover trial, the needle that delivers flagellin across the microbe's membrane corresponds, protein for protein, to the type III secretion system for injecting toxins and other molecules. The resemblance speaks to a common ancestry. The type III secretion system is far from the only structure that is related to parts of flagella. Proteins in the motor, for example, are related to proteins found in other motors that *E. coli* and other bacteria use to pump out molecules from their interior.

Scientists are now developing hypotheses from this evidence to explain how flagella evolved. Pallen and Nicholas Matzke, now a graduate student at the University of California, Berkeley, offered one hypothesis in 2006. Before there were flagella, Pallen and Matzke argued, there were simpler parts carrying out other functions. Gene duplication made extra copies of those parts, and mutations caused the copies to be combined into the evolving flagellum. Today flagella serve one main function: to swim. But their parts did not start out that way.

The flagellum's syringe may have begun as a simple pore that allowed molecules to slip through the inner membrane. A proton-driving motor became linked to it, allowing it to push out big molecules. This primitive system may have allowed ancient bacteria to release signals or toxins. Two kinds of structures eventually evolved from it: the type III secretion system and the needle that injects pieces of the flagellum across the membrane.

The next step in the evolution of flagella may have come when the needle began squirting out sticky proteins. Instead of floating away, these proteins clumped around the pore. Bacteria could have used these sticky proteins as many species do today, to allow them to grip surfaces. The microbes added more proteins to produce hairs, which could reach out farther to find purchase.

In the next step, this sticky hair began to move. A second type of motor became linked to it, which could make the hair quiver. Now the microbe could move. Its crude, random movement may have allowed it to disperse during times of stress. Over time this protoflagellum became fine-tuned. Gene duplication allowed the proteins making up the filament to become a flexible hook at the base and stiff, twisted fibers along the shaft. And finally bacteria began to steer. One of their chemical sensing systems became linked to their flagella, allowing them to change their direction.

This hypothesis is not the unveiling of absolute truth. Scientists don't have that power. What scientists can do is create hypotheses consistent with previous observations—in this case, observations of the variations in flagella, the components that play other roles in bacteria, and the ways in which evolution combines genes for new functions. Pallen and Matzke's hypothesis may well prove to be flawed, but the only way to find out is to search the genomes of *E. coli* and other microbes for more clues as to how the flagellum was assembled, to study how intermediate structures work, and perhaps even to genetically engineer some of the intermediate steps that have disappeared. A better hypothesis may emerge along the way. But it is a far superior hypothesis to one built on nothing but appearances and a personal sense of disbelief.

NETWORKS UNDER CONSTRUCTION

In order to build a flagellum, *E. coli* does not simply churn out all the proteins in a blind rush. It controls the construction with a sophisticated network of genes. Only when it detects signs of stress does it switch on the flagella-building genes, and it uses a noise filter to avoid false alarms. It turns the genes on step by step as it gradually builds up the flagellum, then it turns them off. And like the flagellum, *E. coli*'s control networks have an ancient history of their own.

In 2006, M. Madan Babu, a biologist at the University of Cambridge, and his colleagues published a major investigation of how *E. coli*'s circuitry evolved. They began by searching for *E. coli*'s genetic switches—the proteins that grab on to DNA and turn on, turn off, or otherwise influence other genes. They ended up with more than 250 of them. They then combed through the scientific literature to figure out which genes these switches controlled. All told, Babu and his colleagues mapped a dense web of 1,295 links joining 755 genes.

The map Babu's team drew looks a lot like the hierarchy of a government or a corporation. A few powerful genes sit at the top, each directly controlling several other genes. Those middle-manager genes control many other genes in turn, which may control still others. This organization allows *E. coli* to cope with changes in its environment with swift, massive changes to its biology. Babu's map also let him survey *E. coli*'s network down to its smallest circuits.

Once Babu had finished his map of *E. coli*'s network, he could reconstruct its history. He compared it with the networks in 175 other species of microbes. Babu discovered a network core shared by all of them, made up of 62 genetic switches controlling 376 genes, for a total of 492 links. This core, Babu concluded, existed in the common ancestor of all living things.

This core network offers some hints of what that common ancestor was like. It already had sensors, which allowed it to detect different kinds of sugar and monitor its own energy levels. It could detect oxygen, not to breathe it—since the atmosphere was nearly oxygen free—but probably to protect itself from its own toxic oxygen-bearing waste. This ancestral microbe was already using genetic switches to control iron-scavenging genes, to create the building blocks for proteins and DNA. It was, in other words, a fairly supple little bug.

From that common ancestor every living thing today evolved. Along the way its network evolved as well. The lineage that led to *E. coli* gained new circuits to sense and feed on new sugars, for example. Experiments on living *E. coli* have helped shed light on how mutations and natural selection rewired its network. One of the simplest means by which *E. coli*'s network can be rewired is the accidental duplication of a chunk of DNA. In some cases, the duplication may create two copies of the same switch. If the gene for one of those switches mutates, it may begin to control a different gene. In other cases, extra copies of genes created by duplications are controlled by the switch that turned on the original gene.

The ancestors of *E. coli* rewired their networks as they adapted to new ways of life. Sometimes only minor tinkering with a circuit would produce an important adaptation—adding an extra switch to a gene, for example, or taking one away. One of these tinkered circuits allows *E. coli* to sense a drop in oxygen and switch its metabolism over to oxygen-free pathways. It is almost identical, gene for gene, to an oxygen-sensing circuit in *Haemophilus influenzae,* a species of bacterium that infects the bloodstream. In *H. influenzae* one switch turns on two genes, which then activate all the other genes required to shift the microbe to an oxygen-free metabolism. It's a fast circuit, which suits *H. influenzae* well since it lives in the blood and experiences rapid drops in oxygen as it moves from arteries to veins.

E. coli, on the other hand, does not make snap decisions about oxygen. Living in the relatively stable environment of the gut, it does not experience the sudden, long-term drops in oxygen that *H. influenzae* does. A slight fluctuation might be a false alarm, which would cause *E. coli* to invest a lot of energy making new enzymes that would be of no use. And that fact of life is reflected in *E. coli*'s oxygen circuit. It is identical with *H. influenzae*'s circuit but for one extra gene, called NarL:

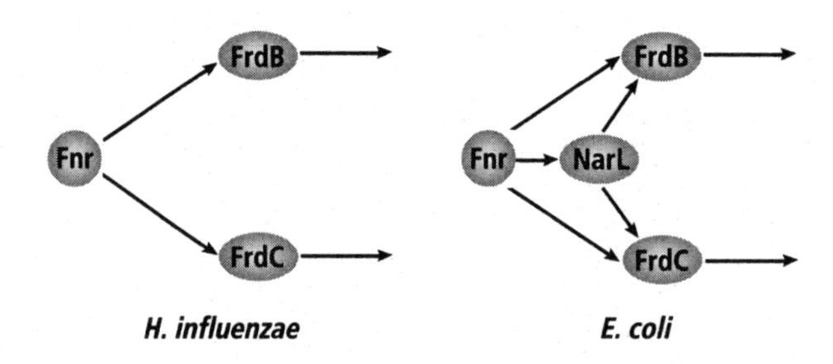

H. influenzae ***E. coli***

In *H. influenzae,* Fnr immediately switches on FrdB and FrdC. But in *E. coli* those genes also need a signal from NarL. It takes time for Fnr to drive the level of NarL high enough to give the two genes both the signals they need. A minor dip in oxygen won't provide them with enough time to prime the pump.

As *E. coli*'s network evolved, it became impressively robust. The growth of man-made networks offers some clues to how that happened. The Internet did not suddenly appear one morning, ready to send your e-mail anywhere in the world. It began in 1969 as a crude link between computers at the University of California in Los Angeles and the Stanford Research Institute in Palo Alto, California. Other institutions joined the network over the years, and more links were added between them. The Internet became robust thanks to its overall architecture. But no one wrote down the design specifications for the entire Internet in 1969. They emerged along the way. Computer engineers focused their attention on how well each small part of the network performed. They worried about the cost of long-range connections between servers, and so they kept the links short.

E. coli's network grew in a similar way. As genes were accidentally duplicated, the network grew more complex. Mutations rewired some of the new genes so that they interacted with other genes. Natural selection then selected the favorable mutations and rejected the rest. As efficient small-scale components evolved, a robust network emerged as a by-product.

At the Dover intelligent design trial, creationists revealed a fondness for analogies to technology. If something in *E. coli* or some other organism looks like a machine, then it must have been designed intelligently. Yet the term *intelligent design* is ultimately an unjustified pat on the back.

The fact that *E. coli* and a man-made network show some striking similarities does not mean *E. coli* was produced by intelligent design. It actually means that human design is a lot less intelligent than we like to think. Instead of some grand, forward-thinking vision, we create some of our greatest inventions through slow, myopic tinkering.

FIRST WORDS

Scrape away *E. coli*'s new genes—the arrivistes carrying resistance to penicillin and other drugs. Peel back the older genes that *E. coli* evolved after splitting off from other bacteria millions of years ago. Strip off the deeper layers, the ones that build *E. coli*'s flagella and the ones that have been destroyed beyond use. Strip away the genes for its peptidoglycan mesh, its sensors for rewards and dangers, its filters and amplifiers. Get rid of the genes that encode the proteins that were carried by the last common ancestor of all living things some 4 billion years ago.

You are not left with a clean sheet. A scattered collection of enigmatic chunks of DNA remains. These are not typical genes. *E. coli* uses them only to make RNA, and that RNA is never used to make proteins. These RNA genes are the oldest level of the palimpsest. Scientists suspect that they are vestiges of some of the earliest organisms that existed on the planet, from a time before DNA.

Life's raw materials are no different from lifeless matter. Stars made the carbon, phosphorus, and other elements in our bodies. If you travel the solar system, you will encounter meteorites and comets with ample supplies of amino acids, formaldehyde, and other compounds found in living things. The Earth incorporated many of these molecules as it formed 4.5 billion years ago, and showers of space dust and the occasional impact of a bigger hunk of rock or ice brought in fresh supplies. The planet acted like a chemical reactor, baking, mixing, and percolating these molecules, probably producing still more molecules essential to life before life yet existed. The great mystery that attracts many scientists is how this reactor gave rise to life as we know it, complete with information-encoding DNA, its single-stranded counterpart RNA, and proteins.

As soon as the basic outlines of molecular biology became clear in the 1960s, scientists decided that DNA, RNA, and proteins did not emerge from the lifeless Earth all at once. But which came first? DNA may be a marvelous repository of information, but without the care provided by proteins and RNA it is just a peculiar string-shaped molecule. Proteins are awesomely versatile, able to snatch atoms drifting by, forge new molecules, and break old ones apart. But they are not so good at storing information for building proteins or for passing that information on to the next generation.

Francis Crick spent many hours in the mid-1960s speculating on the origin of life with his colleague at Cambridge, the chemist Leslie Orgel. They came to the same basic conclusion, one that Carl Woese came to on his own. Perhaps DNA and proteins emerged well after life began on Earth. Perhaps before life depended on DNA and protein, it was based on RNA alone.

At the time the suggestion seemed a little bizarre. RNA's main role in cells appeared to be as a messenger, delivering information from genes to the ribosomes where proteins were made. But Crick, Orgel, and Woese all pointed out that experiments on *E. coli* showed that RNA molecules also have other jobs. The ribosome, for example, is itself made up of dozens of proteins and a few molecules of RNA. Another kind of RNA, called transfer RNA, helps weld amino acids onto the end of a growing protein. Perhaps, the scientists suggested, RNA has a hidden capacity for the sort of chemical acrobatics proteins are so good at. Perhaps RNA was the first molecule to emerge from the lifeless Earth, with different versions of the molecule playing the roles of DNA and protein. Perhaps DNA and proteins evolved later, proving superior at storing information and carrying out chemical reactions, respectively.

Years later Crick and Orgel freely admitted that the idea of primordial RNA went nowhere after they published it in 1968. Fifteen years would pass before people began to take it seriously. A year after Crick proposed an RNA origin for life, a young Canadian biochemist named Sydney Altman arrived at Cambridge to work with him on transfer RNA. Altman discovered that when *E. coli* makes its transfer RNA molecules, it must snip off an extra bit of RNA before they can work properly. Altman named *E. coli*'s snipping enzyme ribonuclease P (RNase P for short). At Cambridge and then at Yale, Altman slowly teased apart RNase P and was surprised to find that it is a chimera: part protein, part RNA. Altman and his colleagues found that the blade that snips the transfer RNA is itself

RNA, not protein. Altman had discovered an RNA molecule behaving like an enzyme—something that had never been reported before.

Altman would share a Nobel Prize in 1989 with Thomas Cech, a biochemist now at the University of Colorado. Cech found similarly strange RNA in a single-celled eukaryote known as *Tetrahymena thermophila,* which lives in ponds. Unlike prokaryotes, eukaryotes must edit out large chunks of RNA interspersed in a gene before they can use it for building proteins. Proteins that build the messenger RNA generally edit out these chunks. But Cech discovered that in *Tetrahymena,* some RNA molecules can splice themselves without any help from a protein. They simply fold precisely back on themselves and cut out their useless parts.

Cech's and Altman's discoveries showed that RNA is far more versatile than anyone had thought. Many biologists turned back to the visionary ideas of Crick, Orgel, and Woese. Perhaps before DNA or proteins evolved, there had existed what Walter Gilbert of Harvard called "the RNA world."

If RNA-based life did once swim the seas, its RNA molecules would have had to be a lot more powerful than the ones discovered by Altman and Cech. Some would have had to serve as genes, able to store information and pass it down to new generations. Others would have had to extract the information in those genes and use it to build other RNA molecules that could act like enzymes. These ribozymes, as they were known, had to capture energy and food and replicate genes.

The possibility of an RNA world spurred a number of scientists to explore the evolutionary potential of this intriguing molecule. In the 1990s, Ronald Breaker, a biochemist at Yale, set out to make RNA-based sensors. He reasoned they would work like the signal detectors found in *E. coli.* They would have to be able to grab particular molecules or atoms, change their shape in response, and then react with other molecules in the microbe.

Breaker didn't design these sensors, though. Instead, he took advantage of the creative powers of evolution. He dumped an assortment of RNA molecules into a flask and then added a particular chemical he wanted his sensor to detect. A few of the RNA molecules bonded clumsily to the chemical while the rest ignored it. Breaker fished out those few good RNA molecules and made new copies of them. He made them sloppily, so that he randomly introduced a few changes to their sequences. In other words, the RNA mutated. When Breaker exposed the mutated RNA molecules to the same chemical again, some of them did an even better job of binding to it. Breaker repeated this cycle of mutation and selection for many rounds, until the RNA molecules could swiftly seize the chemical.

Eventually Breaker and his colleagues were making RNA molecules that could not only grab the chemical but change their shape. These RNA molecules could act like an enzyme, able to cut other RNA molecules in half. Breaker had created an RNA molecule that could sense something in its environment and use the information to do something to other RNA molecules. He dubbed it a riboswitch.

In the years that followed, Breaker created a library of riboswitches. Some can respond precisely to cobalt, others to antibiotics, others to ultraviolet light. RNA's ability to evolve such a range of riboswitches brought more weight to the RNA-world theory. Breaker then had a thought. If the RNA-world theory was right, then RNA-based life had shifted many of the jobs once carried out by RNA to DNA and proteins. But perhaps RNA had not surrendered all those jobs. Perhaps riboswitches still survive in DNA-based organisms. In some cases, an RNA-based sensor might be superior to one made of protein. Riboswitches are easier to make, Breaker noted, since all a cell needs to do is read a gene and make an RNA copy.

Breaker and his students set out on a search for natural riboswitches. In a few months they had found one in *E. coli,* which uses this particular riboswitch to sense vitamin B_{12}. *E. coli* makes its own vitamin B_{12}, which it needs to survive. But above a certain concentration extra B_{12} is just a waste. *E. coli*'s riboswitch, Breaker found, binds vitamin B_{12}. The binding causes it to bend into a shape in which it can shut down the protein that makes the vitamin. Breaker couldn't have fashioned a more elegant riboswitch himself.

Breaker went on to find more riboswitches in *E. coli,* and then he found more in other species. Most of them keep levels of chemicals in balance by swiftly shutting down genes. Since Breaker discovered riboswitches, other scientists have found RNA doing many other things in *E. coli.* Some shut genes off, and others switch them on. Some prevent RNA from being turned into proteins, while others keep its iron in balance. Some RNA molecules allow *E. coli* to communicate with other microbes, and others help it withstand starvation. These RNA molecules form a hidden control network that's only now emerging from the shadows. Their discovery has helped make the RNA world even more persuasive.

Still, the question of exactly how RNA-based life emerged and then gave rise to DNA-based life gives scientists a lot to argue about. Some believe that RNA could have emerged directly from a lifeless Earth. Its ribose backbone, for example, might have been able to form in desert lakes, where borate can keep the fragile sugar stable for decades. Some argue that other replicating molecules came first and that the RNA world was merely one phase of history.

Like any living thing, RNA life needed some kind of boundary. Some scientists argue that RNA organisms did not make their own membranes but, rather, existed in tiny pores of ocean rocks. As RNA molecules replicated, the new copies spread from chamber to chamber. Other scientists see RNA life packaged in more familiar cells. They are trying to create these organisms from scratch, crafting oily bubbles that can trap RNA molecules. Proof by invention is their strategy.

There's probably little to fear from the creation of RNA-based life. Most experts suspect it would survive only in the confines of the laboratory. DNA-based life is far superior in the evolutionary arena. But that doesn't mean DNA-based life has abandoned all the ways of its ancestor. RNA may still work best for certain tasks, and that superiority is why it continues to exert control over *E. coli* and other species. In some ways the RNA world never ended. We still live in it today.

AU REVOIR, MON ÉLÉPHANT

In many ways, Jacques Monod was far more right than he realized when he uttered his famous words about *E. coli* and the elephant. We share with *E. coli* a basic genetic code and many proteins essential for getting energy from food. *E. coli* and our own cells face many of the same challenges. They both need to keep a boundary with the outside world intact yet not too rigid. *E. coli* has to keep its DNA neatly folded and yet accessible for speed-reading. It has to keep track of its inner geography. It needs to organize its thousands of genes into a network that can respond in a coordinated way to changes in the outside world. Its network has to remain rugged and robust despite the fact that it is swamped with noise. *E. coli* communicates with other members of its species, allies with some, fights with others, gives up its life. Like us, it grows old.

Some of these similarities are the result of a common heritage reaching back to the earliest stages of life on Earth. Others are the result of two evolutionary paths that converged on the same solution. Yet even the cases of convergence strengthen Monod's insight. They are evidence that despite 4 billion years of separate history, we and *E. coli* are still deeply sculpted by the same evolutionary forces.

I have met some scientists, however, who simply hate Monod's quip. It tramples over some fundamental differences between the elephant and *E. coli.* Elephants—and humans and lichens and all other eukaryotes—have vastly larger genomes than *E. coli.* Our own genome, for example, has about five times as many genes. It's also padded with a lot of DNA that does not encode proteins. Another major difference can be found in the proteins we use to replicate DNA. They do not show any clear relationship to the proteins used by *E. coli* or other bacteria. Eukaryotes do swap a few genes, but much more rarely than *E. coli* does. We do not shake hands with friends and take up their genes for blue eyes. As animals, we have a way of reproducing that couldn't be more different from *E. coli*'s. Only a tiny fraction of the cells in our bodies have the potential to carry our genes successfully to the next generation, and our genomes carry the information necessary for the stately development of a new trillion-celled body complete with 200 cell types and dozens of organs.

These differences are indeed great and genuine, and yet scientists have surprisingly little idea of how they came to be. Why we're not more like *E. coli* is, in some ways, an open question. The answer must be lurking in the early history of life on Earth. Scientists are agreed that life split into three branches very early on, and the differences among them—particularly those that divide eukaryotes from bacteria and archaea—are profound. Yet at the moment, experts are contemplating some radically different explanations for how those divisions emerged. Some have claimed that eukaryotes originated from archaea that swallowed oxygen-breathing bacteria. Others claim that the split occurred long before that, before life crossed into the DNA world.

I find one explanation particularly intriguing. It comes from Patrick Forterre, an evolutionary biologist at Monod's Pasteur Institute. He proposes that the profound split between us and *E. coli* is the work of viruses.

Forterre's scenario begins in the RNA world, before the three great divisions of life had yet emerged. RNA-based organisms were promiscuously swapping genes. Some of these genes began to specialize, becoming parasites. They no longer built their own gene-replicating machinery but invaded other organisms to use theirs. These were the first viruses, and they are still around us today, in the form of RNA viruses, such as influenza, HIV, and the common cold.

It was these RNA viruses, Forterre argues, that invented DNA. For viruses, DNA might have offered a powerful, immediate benefit. It would have allowed them to ward off attacks by their hosts by combining pairs of single-stranded RNA into double-stranded DNA. The vulnerable bases carrying the virus's genetic information were now nestled on the inside of the double helix while a strong backbone faced outward.

Early DNA viruses probably evolved a range of relationships with their hosts. *E. coli*'s viruses are good to keep in mind here: the lethal ones that make the microbe explode with hundreds of viral offspring, the quiet ones that cause trouble only in times of stress, and the beneficial ones that have become fused seamlessly to their hosts. Forterre argues that on several occasions, DNA viruses became permanently established in their RNA hosts. As they became domesticated, they lost the genes they had used to escape and make protein shells. They became nothing more than naked DNA, encoding genes for their own replication.

Only at that point, Forterre argues, could RNA-based life have made the transition to DNA. From time to time, mutations caused genes from the RNA chromosome to be pasted on the virus's DNA chromosome. The transferred genes could then enjoy all the benefits of DNA-based replication. They were more stable and less prone to devastating mutations. Natural selection favored organisms that carried more genes in DNA than in RNA. Over time, the RNA chromosome shriveled while the DNA chromosome grew. Eventually the organism became completely DNA based. Even the genes for riboswitches and other relics of the RNA world were converted to DNA. Forterre proposes that this viral takeover occurred three times. Each infection gave rise to one of the three domains of life.

Forterre argues that his scenario can account for the deep discord between the genes that all three domains share and the ones that are different. Forterre started his scientific career studying the enzymes *E. coli* uses to build DNA. Related versions of those enzymes exist in other species of bacteria, but they are nowhere to be found in archaea or eukaryotes. The difference, Forterre argues, lies in the fact that the ancestors of *E. coli* and other bacteria got their DNA-building enzymes from one strain of virus and the eukaryotes and archaea didn't.

Once the three domains split, they followed different trajectories. Our own ancestors, the early eukaryotes, may have acquired their nucleus and other traits from other viruses. Eukaryotes grew to be larger than bacteria or archaea, and as a result their populations grew smaller. In small populations it's easier for slightly harmful mutations to spread, thanks merely to chance. It may have been only then that the eukaryote genome began to expand. Interspersing noncoding DNA within genes may have been harmful at first, but over time it may have given eukaryotes the ability to shuffle segments of their genes to encode different proteins. We humans have 18,000 genes, but we can make 100,000 proteins out of them.

Forterre's proposal is as radical as the suggestion in 1968 that life was once based on RNA. It will demand just as much research to test. In the meantime, it is intriguing to think about what it would mean if Forterre is right. The differences between the elephant and *E. coli* would actually be the sign of yet another fundamental similarity: we—all living things—are different only because we got sick from different viruses.

Ten

PLAYING NATURE

PORTRAIT IN PROTOPLASM

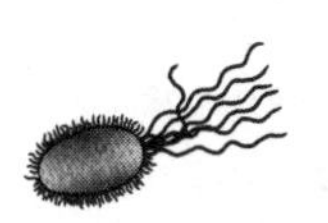

IN CHRISTOPHER VOIGT'S LABORATORY at the University of California, San Francisco, you can have your picture taken by *E. coli.* Voigt will place a photograph of you before a hooded contraption. The reflected light from the picture strikes a tray covered with a thin, gummy layer of *E. coli.* It's a special strain that Voigt and his colleagues created in 2005. They inserted genes into the bacteria, some of which let the bacteria detect light and some of which cause them to produce a dark pigment. The genes are wired so that if a microbe detects light—such as the light reflected from a photograph—it shuts down the genes for making pigment. The bacteria that catch photons from light parts of the picture remain clear. The ones that don't churn out pigment and turn sepia. A picture emerges, soft, fuzzy, but recognizably you.

Voigt is an assistant professor with a long list of scientific papers on his résumé. But he is also a child of the biotechnological age. He had not yet been born when scientists first learned how to insert genes in *E. coli* in the 1970s. That breakthrough was one of the most important in the history of biology. Genetic engineering allowed scientists to decipher some of the genome's most baffling features. They turned *E. coli* into an industrial workhorse and created a $75 billion industry. Once scientists had mastered the art of inserting genes into *E. coli,* they began putting them in other microbes and then in animals and plants. Now goats produce drugs in their milk. Now 250 million acres of farmland are covered in crops carrying genes that make them resistant to pesticides and herbicides.

But as genetic engineering spreads to other species, *E. coli* has not faded into the background. It remains the species of choice for scientists who want to develop new tools for manipulating life. Voigt's work, for example, is part of a new kind of genetic engineering called synthetic biology. Instead of simply moving a single gene from one species to another, synthetic biologists seek to create entire circuits of genes. They wire together genes from various species and fine-tune them to carry out new functions. For now synthetic biologists have learned enough only to create eye-catching proofs of principle, like Voigt's microbial camera. But these lessons could lead to microbes that act as solar-power generators, or that can produce drugs when the conditions are right—call them thinking drugs. Some synthetic biologists are even trying to dismantle *E. coli* and use its parts to rebuild life from scratch.

This new research tingles with controversy. A debate is raging over the risks posed by synthetic biology and other advances in biotechnology—the accidental release of dangerous new creatures, for example, or even intentional engineering of biological weapons. Thinking drugs could become thinking plagues. Synthetic biologists have also given a fresh spur to the debate over the morality of biotechnology in general. Today the world faces a huge, confusing surge of scientific research, with mice growing human neurons in their brains and deadly viruses being built from the ground up. In order to resolve these debates, we must think seriously about what it means to be alive and how biotechnology changes that meaning. And *E. coli*, the germ of our biotechnological age, has much to tell us. The face looks back, less a portrait than a mirror.

NEOLITHIC BIOTECH

Biotechnology was born many times, and each time it was born blind.

Humans began to manipulate other life-forms to make useful things, such as food and clothing, at least 10,000 years ago. In places such as Southeast Asia, Turkey, West Africa, and Mexico, people began to domesticate animals and plants. They probably did so unwittingly at first. Gathering plants, they picked some kinds over others, accidentally spreading the seeds on the ground. The wild ancestors of dogs that lingered near campfires might have fed on scraps and passed on their sociable genes to their pups. These species adapted to life with humans through natural selection. Once humans began to farm and raise livestock, natural selection gave way to artificial selection as they consciously chose the individuals with the traits they wanted to breed. Evolution accelerated as humans assembled a parade of grotesque creations, from flat-faced pugs to boulder-sized pumpkins.

The first Neolithic biotechnologists were manipulating microbes as well. They learned how to make beer and wine or, rather, how to allow yeast to make beer and wine. The job of humans was simply to create the best conditions in which the yeast could transform sugar to alcohol. Yeast also lifted bread with its puffs of carbon dioxide. Domesticated microbes evolved just as weedy teosinte evolved into corn and scrawny jungle fowls evolved into chickens. The yeast of winemakers became distinct from its wild cousins that still lived on tree bark.

With the invention of yogurt an entire ecosystem of bacteria evolved. Yogurt was first developed by nomadic herders in the Near East about five thousand years ago. They probably happened to notice one day that some milk had turned thick and tangy, and that it also proved slow to turn rancid. Plant-feeding bacteria had fallen into the milk and had altered its chemistry as they fed on it. The herders found that adding some of the yogurt to normal milk transformed it into yogurt as well. The bacteria in those cultures became trapped in a new ecosystem, and they adapted to it, evolving into better milk feeders and jettisoning many of the genes they no longer needed.

For thousands of years, humans continued to tinker with animals, plants, and microbes in this same semiconscious way. But as the microbial world unfolded beginning in the nineteenth century, scientists discovered new ways to manipulate nature. The first attempts were simple yet powerful. Louis Pasteur demonstrated that bacteria turned wine sour and contaminated milk. Heat killed off these harmful microbes, leaving children healthier and oenophiles happier.

As microbiologists discovered microbial alchemy, they searched for species that could carry out new kinds of useful chemistry. Chaim Weizmann, the first president of Israel, originally came to fame through his work in biotechnology. Living in Britain during World War I, he discovered bacteria that could manufacture acetone, an ingredient in explosives. Winston Churchill quickly took advantage of it by building a string of factories to breed the bacteria in order to make cheap acetone for the Royal Navy. The next generation of microbiologists began manipulating genes to make them even more efficient. By bombarding the mold that makes penicillin, scientists created mutants with extra copies of penicillin genes, allowing the mold to make more of the drug.

As scientists discovered how to manipulate life, they wondered what sort of world they were creating. In a 1923 essay, the British biologist J.B.S. Haldane indulged in some science fiction. He pretended to be a historian of the future looking back on the 1940 creation of a new strain of algae that could pull nitrogen from the air. Strewn on crops, it fertilized them so effectively that it doubled the yield of wheat. But some of the algae escaped to the sea, where it turned the Atlantic to jelly. Eventually it triggered an explosion in the population of fish, enough to feed all humanity.

"It was of course as a result of its invasion by *Porphyrococcus* that the sea assumed the intense purple colour which seems so natural to us, but which so distressed the more aesthetically minded of our great grandparents who witnessed the change," Haldane wrote. "It is certainly curious to us to read of the sea as having been green or blue."

For the next fifty years, hope and dread continued to tug scientists in opposite directions. Some hoped that biotechnology would offer an alternative to a polluted nuclear-powered modern world, a utopia in which poor nations could find food and health without destroying their natural resources. Yet the notion of rewriting the recipe for life sometimes inspired disgust rather than wonder. It might well be possible to create an edible strain of yeast that could feed on oil. But who would want to eat it?

Aside from scientists, few people took these speculations very seriously. For all the progress biotechnology made up until 1970, there was no sign that life would change anytime soon. And then, quite suddenly, scientists realized they had the power to tinker with the genetic code. They could create a chimera with genes from different species. And they began their transformation of life with *E. coli.* Monod's motto took on yet another meaning: if scientists could genetically engineer *E. coli,* there was every reason to believe they would someday engineer elephants.

CUT AND PASTE

Before 1970, *E. coli* had no role in biotechnology. It does not naturally produce penicillin or any other precious molecule. It does not turn barley into beer. Most scientists who studied *E. coli* before 1970 did so to understand how life works, not to learn how to make a profit. They learned a great deal about how *E. coli* uses genes to build proteins, how those genes are switched on and off, how its proteins help make its life possible. But in order to learn how *E. coli* lives, they had to build tools to manipulate it. And those tools would eventually be used to manipulate *E. coli* not simply to learn about life but to make fortunes.

The potential for genetic engineering took *E. coli*'s biologists almost by surprise. In the late 1960s, a Harvard biologist named Jonathan Beckwith was studying the *lac* operon, the set of genes that *E. coli* switches on to feed on lactose. To understand the nature of its switch, Beckwith decided to snip the operon out of *E. coli*'s chromosome. He took advantage of the fact that some viruses that infect the bacteria can accidentally copy the *lac* operon along with their own genes. Beckwith and his colleagues separated the twin strands of the DNA from two different viruses. The strands containing the *lac* operon had matching sequences, so they were able to rejoin themselves. Beckwith and his colleagues added chemicals to the viruses that destroyed single-strand DNA, leaving behind only the double-strand operon. For the first time in history someone had isolated genes.

On November 22, 1969, Beckwith met the press to announce the discovery. He let the world know he was deeply disturbed by what he had just done. If he could isolate genes from *E. coli,* how long would it take for someone else to figure out a sinister twist on his methods—a way to create a new plague or to engineer new kinds of human beings? "The steps do not exist now," he said, "but it is not inconceivable that within not too long it could be used, and it becomes more and more frightening—especially when we see work in biology used by our Government in Vietnam and in devising chemical and biological weapons."

Beckwith flashed across the front page of *The New York Times* and other newspapers, and then he was gone. The debate over the dangers of genetic engineering disappeared. Other scientists went on searching for new ways to manipulate genes without giving much thought to the danger. Scientists who studied human biology looked jealously at the tools Beckwith and others could use on *E. coli.* To study a single mouse gene, a scientist might need the DNA from hundreds of thousands of mice. As a result, they knew very little about how animal cells translated genes into proteins. They knew even less about the genes themselves—how many genes humans carry, for example, or the function of each one.

Paul Berg, a scientist at Stanford University, spent many years studying how *E. coli* builds molecules, and in the late 1960s he wondered if he could study animal cells in the same way. At the time, scientists were learning about a new kind of virus that permanently inserts itself into the chromosomes of animals. The virus was medically important because it could cause its host cells to replicate uncontrollably and form tumors. Berg recognized a similarity between these animal viruses and some of the viruses that infect *E. coli.* In the 1950s, scientists had learned how to turn *E. coli*'s viruses into ferries to carry genes from one host to another. Berg wanted to know whether animal viruses could be ferries as well.

Berg began to experiment with a cancer-causing monkey virus called SV40. He pondered how he might insert another gene into it. Eventually he decided he would need to cut open the circular chromosome of SV40 at a specific point. But he had no molecular knife that could make that particular cut.

As it happened, other scientists had just found the knife. In the 1960s, scientists had discovered *E. coli*'s restriction enzymes, which slice up foreign DNA by grabbing on to certain short sequences. One of those scientists was Herbert Boyer, a microbiologist at the University of California, San Francisco. Boyer gave Berg a supply of a restriction enzyme he had recently discovered, called EcoR1.

Berg and his colleagues used EcoR1 to cut open SV40's chromosome. At one end of SV40's DNA they added DNA from a virus of *E. coli* called lambda. In order to fuse the two pieces of DNA together, Berg and his colleagues added to their ends some extra bases that would form bonds. When they were done, they had created a viral hybrid.

Since the hybrid carried the lambda virus's genes for invading *E. coli,* Berg wondered whether it could invade the microbe. He asked one of his graduate students, Janet Mertz, to design an experiment. For Berg and Mertz, the experiment started out as yet another interesting question. But some who learned about their plans were filled with dread.

One of the first people to confront Berg with these worries was a bioethicist named Leon Kass. Like Berg, Kass had worked on *E. coli,* but he had become disillusioned by how fast scientific discoveries were being made and the lack of thought being given to their ethics. Kass warned

Berg that manipulating genes could lead to moral quandaries. If scientists could insert genes in embryos, parents might pick out the traits they wanted in their children. They wouldn't just upgrade genes that would cause sickle-cell anemia or other genetic disorders. They would look for ways to enhance even perfectly healthy children.

"Are we wise enough to be tampering with the balance of the gene pool?" Kass asked Berg.

Berg brushed off Kass's warning, but when other virus experts began to question his plans, he stopped short. Mertz described to another researcher how she and Berg were going to create a sort of Russian doll with SV40 in lambda and lambda in *E. coli.* The researcher replied, "Well, it's *coli* in people."

If an SV40-carrying *E. coli* escaped from Berg's laboratory, some scientists feared it might make its way into a human host. Once inside a person, it might multiply, spreading its cancer-causing viruses. No one could say whether it would do no harm or trigger a cancer epidemic. In the face of these uncertainties, Berg and Mertz decided to abandon the experiment.

"I didn't want to be the person who went ahead and created a monster that killed a million people," Mertz said later.

At the time, Berg's lab was the only one in the world actively trying to do genetic engineering. The researchers' methods were elaborate, tedious, and time-consuming. When they scrapped their SV40 experiment, they could be confident that no one would be able to immediately take up where they left off. But it would not be long before genetic engineering would become far easier—and thus far more controversial.

Berg and Boyer continued to study how EcoR1 cuts DNA. They discovered that the enzyme does not make a clean slice. Instead, it leaves ragged fragments, with one strand of DNA extending farther than the other at each end. That dangling strand can spontaneously join another dangling strand also cut by EcoR1. The strands are, in essence, sticky. Berg and Boyer realized no tedious tacking on of extra DNA was necessary to join two pieces of DNA from different species. The molecules would do the hard work on their own.

Boyer soon took advantage of these sticky ends. Instead of viruses, he chose plasmids, those ringlets of DNA that bacteria trade. Working with the plasmid expert Stanley Cohen, Boyer cut apart two plasmids with EcoR1. Their sticky ends joined together, combining the plasmids into a single loop. Each plasmid carried genes that provided resistance to a different antibiotic, and when Boyer and Cohen inserted their new hybrid plasmid in *E. coli,* the bacteria could resist both drugs. And when one of these engineered microbes divided, the two new *E. coli* also carried the same engineered plasmids. For the first time a living microbe carried genes intentionally combined by humans.

Once Boyer and Cohen had combined two *E. coli* plasmids, they turned to another species. Working with John Morrow of Stanford University, they cut up fragments of DNA from an African clawed frog and inserted it in a plasmid, which they then inserted in *E. coli.* Now they had created a chimera that was part *E. coli,* part animal.

When Boyer described his chimeras at a conference in New Hampshire in 1973, the audience of scientists was shocked. None of them could say the experiments were safe. They sent a letter to the National Academy of Sciences to express their concern, and a conversation spread through scientific circles. What could scientists realistically hope to do with engineered *E. coli*? What were the plausible risks?

The possibilities sounded as outlandish as anything Haldane had dreamed of fifty years earlier. *E. coli* could make precious molecules, such as human insulin, which could treat diabetes. *E. coli* might acquire genes for breaking down cellulose, the tough fibers in plants. A person who swallowed cellulose-eating *E. coli* might be able to live on grass. Or maybe engineering *E. coli* would lead to disaster. A cellulose-digesting microbe might cause people to absorb too many calories and become hideously obese. Or perhaps it might rob people of the benefits of undigested roughage—including, perhaps, protection from cancer.

Paul Berg and thirteen other prominent scientists wrote a letter to the National Academy of Sciences in 1974 calling for a moratorium on transferred genes—also known as recombinant DNA—until scientists could agree on some guidelines. The first pass at those guidelines emerged from a meeting Berg organized in February 1975 at the Asilomar Conference Grounds on the California coast. Rather than calling for an outright ban on genetic engineering, the scientists advocated a ladder of increasingly strict controls. The greater the chance an experiment might cause harm, the more care scientists should take to prevent engineered organisms from escaping. Some particularly dangerous experiments, such as shuttling genes for powerful toxins into new hosts, ought not to be carried out at all. The National Institutes of Health followed up on the Asilomar meeting by forming a committee to set up official guidelines later that year.

To scientists such as Berg, these steps seemed reasonable. They had taken time to give genetic engineering some serious reflection, and they had decided that its risks could be managed. Genetic engineering was unlikely to trigger a new cancer epidemic, for example, because from childhood on people were already exposed to cancer-causing viruses. Many scientists concluded that *E. coli* K-12 had become so feeble after decades of laboratory luxury that it probably could not survive in the human gut. A biologist named H. William Smith announced at Asilomar that he had drunk a solution of *E. coli* K-12 and found no trace of it in his stool. But to be even more certain that no danger would come from genetic engineering, Roy Curtiss, a University of Alabama microbiologist, created a superfeeble strain that was a hundred million times weaker than K-12.

Other scientists did not feel as confident. Liebe Cavalieri, a biochemist at the Sloan-Kettering Institute in New York, published an essay in *The New York Times Magazine* called "New Strains of Life—or Death." Below the headline was a giant portrait of *E. coli* embracing one another with their slender alien pili. Meet your new Frankenstein.

Soon the scientific critics were joined by politicians and activists. Congress opened hearings on genetic engineering, and representatives introduced a dozen bills calling for various levels of control. City politicians took action as well. The mayor of Cambridge, Massachusetts, Alfred Vellucci, held raucous hearings on Harvard's entry into the genetic engineering game. The city banned genetic engineering altogether for months. Protesters waved signs at scientific conferences, and environmental groups filed lawsuits against the National Institutes of Health, accusing it of not looking into the environmental risks of genetic engineering.

Many critics were appalled that scientists would presume to judge how to handle the risks of genetic engineering on their own. "It was never the intention of those who might be called the Molecular Biology Establishment to take this issue to the general public to decide," James Watson wrote frankly in 1981. The critics argued that the public had a right to decide how to manage the risk of genetic engineering because the public would have to cope with any harm that might come of it. Senator Edward

Kennedy of Massachusetts complained that "scientists alone decided to impose a moratorium, and scientists alone decided to lift it."

Some critics also questioned whether scientists could be objective about genetic engineering. It was in their interest to keep regulations as lax as possible because they would be able to get more research done in less time. "The lure of the Nobel Prize is a strong force motivating scientists in the field," Cavalieri warned. Along with scientific glory came the prospect of riches. Corporations and investors were beginning to court molecular biologists, hoping to find commercial applications for genetic engineering. Financial interests might lead some to oversell the promise of genetic engineering and downplay its risks. Cetus Corporation, a company that recruited molecular biologists to serve on its board, made this astonishing prediction: "By the year 2000 virtually all the major human diseases will regularly succumb to treatment by disease-specific artificial proteins produced by specialized hybrid micro-organisms."

Instead of a miracle, critics saw in genetic engineering the illusion of a quick fix. In 1977, the National Academy of Sciences held a public forum on the risks and benefits of the new technology. Picketers tried to shut down the meeting, calling genetic engineers Nazis. Amid the chaos, Irving Johnson, the vice president of research at Eli Lilly, talked about how genetic engineering could be used to treat diabetes. Eli Lilly, the country's biggest provider of insulin, got the hormone from the pancreases of pigs. That supply was vulnerable, Johnson said, to a slump in the pork business or to an increase in the population of diabetics. Genetically engineering a microbe to make human insulin might provide a vast, cheap supply. "This is truly 'science for the people,' " Johnson said.

Ruth Hubbard, a Harvard biologist and a leading critic of genetic engineering, testified against this sunny view. She pointed out that insulin does not prevent diabetes or even cure it. It merely counteracts some of the symptoms of the disease. "Before we jump at technological gimmicks to cure complicated diseases," she warned, "we first have to know what causes the diseases, we have to know how the therapy that we are being told is needed works, we have to know what fraction of people really need it. . . . But what we don't need right now is a new, potentially hazardous technology for producing insulin that will profit only the people who are producing it."

While genetic engineering was distracting society from real solutions, critics warned, it could also put the world at risk. What made it particularly risky was its utter dependence on *E. coli.* "From the point of public health," Cavalieri declared, "this bacterium is the worst of all possible choices. It is a normal inhabitant of the human digestive tract and can easily enter the body through the mouth or nose. Once there, it can multiply and remain permanently. Thus every laboratory working with *E. coli* recombinants is staffed by potential carriers who could spread a dangerous recombinant to the rest of the world."

Even if scientists used a weakened strain of *E. coli* for genetic engineering, the microbes might survive long enough outside a lab to pass their engineered genes to more rugged strains. Critics warned of cancer epidemics caused by *E. coli* casually poured down a drain. *E. coli* might churn out insulin inside diabetics, sending them into comas. Genetically engineered organisms could cause bigger disasters than toxic chemicals because they had the reproductive power of life. Erwin Chargaff, an eminent Columbia University biologist, called genetic engineering "an irreversible attack on the biosphere."

"The world is given to us on loan," Chargaff warned. "We come and we go; and after a time we leave earth and air and water to others who come after us. My generation, or perhaps the one preceding mine, has been the first to engage, under the leadership of the exact sciences, in a destructive colonial warfare against nature. The future will curse us for it."

These attacks left the champions of genetic engineering stunned. The debate had become "nightmarish and disastrous," Paul Berg declared in 1979. Stanley Cohen called it a "breeding ground for a horde of publicists."

James Watson, as usual, was bluntest of all. "We were jackasses," he said, looking back at his support of the 1974 moratorium. "It was a decision I regret; one that I am intellectually ashamed of." It had led the public to distract itself from real threats with illusions of apocalypse.

"I'm afraid that by crying wolf about dangers which we have no reason at all to worry about, we are becoming indistinguishable from my two small boys," he wrote. "They love to talk about monsters because they know they will never meet one."

E. COLI, INC.

One figure noticeably absent from the debate was Herbert Boyer, the scientist who had triggered the genetic engineering controversy in the first place. He was busy hunting for companies and investors who could help him make money from his restriction enzymes. In 1976, he became a partner with a young entrepreneur named Robert Swanson. Each man ponied up $500 to launch a company they called Genentech (short for genetic engineering technology). Boyer had to borrow his share.

Boyer and Swanson set out to sell valuable molecules produced by engineered *E. coli.* They decided their first goal should be human insulin, for many of the reasons Irving Johnson had offered to the National Academy of Sciences. Boyer turned to Arthur Riggs and Keiichi Itakura at the City of Hope Hospital in Duarte, California, for help. Riggs and Itakura were among the first scientists learning how to build genes from scratch. When Boyer contacted them, they were in the midst of synthesizing their first human gene, which encoded the hormone somatostatin. Working with Genentech, Riggs and Itakura figured out how to add sticky ends to an artificial somatostatin gene and insert it into a plasmid. They put the plasmid in *E. coli,* which then began to produce somatostatin. It was yet another milestone in a very young science. In 1973, Boyer, Cohen, and Morrow had managed only to put a fragment of an animal gene in *E. coli.* Four years later, Genentech had *E. coli* that could make human proteins.

The scientists did not take long to savor the glory. After they announced their results in 1977, they moved on to insulin. Boyer knew he would have to move fast. Walter Gilbert, the brilliant Harvard molecular biologist, was trying to make insulin as well. But Boyer had a crucial advantage over Gilbert: Boyer's DNA was artificial. Gilbert was trying to isolate insulin DNA from real cells, so his research was subject to the tight grip of government regulation. His team had to take extraordinary precautions and even flew to England to work in a lab set up for biological warfare research. Boyer could move faster because his DNA was not "natural." Instead of isolating it from a cell as Gilbert was doing, Riggs and Itakura worked their way backward from the insulin protein to the sequence of the insulin's gene. Free of regulations, Boyer won the race. On September 6, 1978, Genentech announced that its scientists had extracted 20 billionths of a gram of human insulin from *E. coli.*

Over the next two years, Genentech researchers boosted the yield. They engineered *E. coli* so that it would push its insulin out of its membrane, making it easier to harvest. In 1980, Genentech was ready to hand over the production of insulin to Eli Lilly. The following year the pharmaceutical giant built 40,000-liter tanks, in which it began to breed *E. coli.* Genentech went public, and Boyer's $500 became $66 million.

As Genentech's fortunes waxed, the controversy over *E. coli* waned. Congress never passed a genetic engineering bill, thanks in part to fierce lobbying by scientists. The National Institutes of Health relaxed its guidelines. Scientists working on *E. coli* no longer had to dress up in space suits. Corporations snatched up *E. coli* experts in increasing numbers. All fourteen signatories to Paul Berg's original moratorium letter ended up associated with one venture or another. Walter Gilbert helped launch a company called Biogen, which began engineering *E. coli* to spew out proteins that showed promise of fighting cancer. When Biogen opened its headquarters in Cambridge, Gilbert's old nemesis, former mayor Alfred Vellucci, was there to cut the ribbon.

Genentech led the way for the new biotechnology. Humulin, its microbe-produced insulin, went on the market in 1983, and now 4 million people worldwide take the drug. Other companies make their own brands of *E. coli*–produced insulin, which are used by millions of other diabetics. Biotechnology firms have developed many other drugs from *E. coli,* ranging from human growth hormone to blood thinners. Today *E. coli* churns out vitamins and amino acids. Traditionally, cheese is made by spiking milk with rennet, an enzyme produced in cows' stomachs. Now much of the cheese in stores is made with rennet produced by *E. coli.* Scientists are adding new genes to *E. coli* to see what sorts of new things they can produce, from biodegradable plastics to gasoline.

These advances have not come easily. Scientists cannot simply treat *E. coli* as a machine. The microbe is a living thing, and it responds to manipulation in unexpected ways. Packing the bacteria in a giant tank can cause them to suffocate in their own waste. Engineering them to produce huge amounts of insulin or some other foreign protein puts them under tremendous stress. If the proteins clump together, *E. coli* produces heat-shock proteins to try to untangle them. All the energy *E. coli* uses up coping with the stress is energy it cannot use to feed and grow. Scientists, like cooks perfecting recipes, have struggled to find solutions to these quandaries.

Thirty years have now passed since *E. coli* became the monster and the mule of genetic engineering. It remains one of biotechnology's favorite microbes. Scientists continue to experiment on it to find new ways to manipulate genes and proteins. Its restriction enzymes are the blade of choice for slicing DNA, and its plasmids are the favored breeders of new copies of genes. But scientists can now insert these genes in many other species as well. In the 1980s, they began using the lessons they learned from *E. coli* to shuttle genes into other bacteria and into fungi. Scientists have also learned how to introduce genes into animal and plant cells. Paul Berg's original dream has become real: it is now possible to load a gene on a virus such as SV40 and infect an isolated mammal cell. (Cells from the ovaries of Chinese hamsters are a popular choice.) An engineered cell can then multiply into a laboratory colony, which can then churn out a valuable protein.

It's now also possible to inject genes into living plants. Genetically modified crops now grow across vast stretches of farmland in many countries. Some crops produce a toxin normally made by bacteria that kills insects. Others have been engineered to withstand a weed killer. Scientists have also succeeded in creating plants that can produce human antibodies and vaccines.

Even animals now acquire foreign genes from engineered viruses. Some researchers hope they will be able to treat genetic disorders by supplying cells with working copies of essential genes. Others are inserting genes directly into embryonic cells to produce animals with foreign genes throughout their bodies. Some scientists are trying to ease the pollution produced by farms with this sort of genetic engineering. One major form of pollution from farms is the phosphates—compounds of phosphorus and oxygen—concentrated in fertilizer. When fertilizer washes out into rivers and oceans, the phosphates cause algae blooms and other ecological upheavals that eventually create vast dead zones where nothing can survive. One reason for the high levels of phosphates in fertilizer is that much of it comes from the manure of livestock such as pigs and chickens. These animals cannot break down the phosphates in their food, so it just goes straight through their digestive systems. *E. coli,* among other bacteria, make enzymes that can break down those phosphate-bearing molecules. When researchers insert *E. coli*'s genes in pigs, the animals produce manure that has only a quarter of the normal level of phosphates.

It's chimeric turnaround: thirty years ago scientists were putting animal genes into *E. coli.* Now they are giving animals the genes of *E. coli.*

EXPANDING LIFE'S ALPHABET

Herbert Boyer used his intimate knowledge of *E. coli*'s biology to help create genetic engineering. Today scientists are using his tools to learn more about *E. coli* itself. In the process, they're answering some of the most fundamental questions about life.

Scientists have long debated why life on Earth, with almost no exception, uses only twenty amino acids to build proteins. (*E. coli* is unusual in its ability to make a twenty-first amino acid, called selenocysteine.) There are hundreds of perfectly respectable kinds of amino acids life could have chosen from. To join the Amino Acid Club, a molecule needs only the proper ends. It must have a cluster of nitrogen and hydrogen atoms at one end (an amine) and a cluster of carbon, hydrogen, and oxygen on the other (a carboxyl group). An amine from one amino acid snaps onto the carboxyl group of another like LEGO pieces. It matters little what lies in between. A chemist can synthesize hundreds of different amino acids, and so can the chemistry of outer space. In 1969, a meteorite coated with tarry goo fell to Earth. Scientists found seventy-nine kinds of amino acids lurking inside it.

So why do we have just twenty? One way to investigate the question is to try to produce an organism that can make more. In 2001, Peter G. Schultz of Scripps Research Institute in La Jolla, California, and his colleagues did just that, by engineering *E. coli.* Like other living things, *E. coli* uses a genetic code in which three bases of DNA translate into one amino acid. There are sixty-four possible codons in *E. coli*'s genetic code, most of which it uses regularly. Schultz and his colleagues identified one that it uses only rarely. They engineered *E. coli* so that this neglected codon now instructed the microbe to add an unnatural amino acid to a protein.

Science magazine hailed the achievement as "the first synthetic life form with a chemistry unlike anything found in nature." In the years since, scientists have added over thirty more unnatural acids to *E. coli*'s repertoire. Originally *E. coli* could make these new proteins only if it was supplied with the unnatural amino acids. Recently scientists have begun engineering *E. coli* to make unnatural amino acids from its natural food.

This research has pushed the debate over the genetic code to new ground. No one can argue that life's twenty amino acids are the only ones that can make life possible. Some scientists now argue that the genetic code is just a historical artifact. Early life built its proteins with the most abundant amino acids on the planet, and that unconscious choice was frozen in place. Other scientists argue that the genetic code is actually the best of all possible codes. It offers the biggest range of potential proteins with the fewest genes. And still other scientists argue that natural selection produced the genetic code because it is robust, with the least risk of producing a lethally deformed protein if a mutation strikes a gene.

In our hands, however, the rules of the genetic code have changed. Schultz and other researchers are looking for practical applications for *E. coli*'s unnatural proteins. Unnatural proteins may allow *E. coli* to overcome one of genetic engineering's biggest failures. Unlike bacteria, human cells decorate many of their proteins with knobs of sugar. The sugars force the proteins into new shapes, allowing them to take on new functions. *E. coli* can make perfect copies of our proteins, amino acid for amino acid, but if it can't add the sugars, many of its proteins are useless to us.

Schultz and his colleagues have found a way around this shortcoming. Instead of adding the sugar after a protein is built, they add it to individual amino acids. They then engineer *E. coli* to recognize the unnatural sugarcoated amino acids instead of the ones it normally uses. In this arrangement the bacteria can assemble proteins with sugar knobs already in place, ready for human consumption. What is unnatural for *E. coli* turns out to be quite natural for us.

NETWORK HACKS

For all the futuristic aura around genetic engineering, the science is rather quaint. It is based on a 1950s view of biology. In the world of genetic engineering, *E. coli* and other species are nothing more than simple chemical factories manufacturing their own sets of proteins. Change a gene and you change one of the proteins that comes out. Genetic engineers are well aware that there is much more to life than the production of proteins. There are repressors and promoters, for example, which turn genes on and off. But many genetic engineers use these insights only to make *E. coli* and other organisms into even better factories.

There's another way to look at *E. coli:* as a network. Its proteins and genes work together, allowing the microbe to process information, to make decisions, to keep its biology steady in an unsteady world. The powers of this network emerge from the sum of its parts, not from any one gene or protein. Engineers regularly improve on man-made networks—rewiring circuits, swapping parts. If life follows engineering principles as well, some scientists wonder, would it be possible to rewire life, too?

The first two reports of rewired life came in 2000, and in both cases the life in question was *E. coli.* Michael Elowitz at California Institute of Technology and Stanislas Leibler of Rockefeller University in New York made the microbe blink. They used three genes to build a circuit. Each gene made a different repressor. Elowitz and Leibler engineered the first gene so that its repressor shut down the second gene. The repressor made from the second gene shut down the third. The third shut down the first, but it also did something else: it caused *E. coli* to build a glowing-jellyfish protein.

Elowitz and Leibler found that in some of their engineered microbes, the three repressors became locked in a cycle. As the first gene made more and more repressors, it shut down the activity of the second gene, freeing the third gene to shut down the first one. As the first one stopped making its repressor, the second gene was freed and shut down the third gene, and so on. Elowitz and Leibler arranged these genes on a plasmid and inserted them in *E. coli.* As the genes became active, the scientists could witness this cycle with their own eyes: as the third gene switched on and off, it produced more and then less light. In other words, *E. coli* blinked.

The second report came from the laboratory of James Collins at Boston University. Collins and his colleagues gave *E. coli* a toggle switch. They built two genes, each encoding a repressor that shut off the other gene. Each repressor could be pulled off *E. coli*'s DNA by adding a different molecule to the microbe. To observe how this new circuit of genes worked, Collins and his colleagues, like Elowitz and Leibler, added instructions to one of the genes for building a glowing protein. Adding one kind of molecule caused *E. coli* to start glowing and to continue glowing even after the molecule had run out. Adding the other kind of molecule shut the glow down and kept *E. coli* dark even after it, too, had run out.

These experiments are now recognized as marking the birth of a synthetic biology. It was a humble start, when you consider that a clever child with a home electronics kit can make a blinking light or a toggle switch. But once biologists and engineers learn how to make simple genetic circuits, they move on to complex ones. Combine some simple logic gates and you can end up with a powerful computer chip.

I am writing this book only seven years after the birth of synthetic biology, and scientists are still a long way from building *E. coli* with the equivalent of a computer chip inside. But they have come a long way from toggle switches and blinkers. The *E. coli* camera is a good example of what they can do now. Each year Massachusetts Institute of Technology hosts a synthetic-biology tournament, in which students try to transform *E. coli* into various devices. In 2004, students at the University of Texas and the University of California, San Francisco, worked together to make bacteria that could capture an image. They envisioned a film of engineered *E. coli* that would behave like a piece of traditional photographic film. The bacteria would turn dark unless they were struck by light. The more light that struck them, the less dark they would become. Normally, *E. coli* cannot sense light, nor can it produce colors. But the students were able to engineer a strain that does both. They borrowed a gene for a light-sensitive receptor from a species of blue-green alga called *Synechocystis.* To color the microbes, they borrowed genes from *Synechocystis* that create pigments.

The hard part of the work came when it was time to join the two sets of genes. The students engineered the light receptors so that they could pass a signal to molecules normally made by *E. coli.* Those molecules were then able to grab on to the microbe's DNA and shut down the production of *Synechocystis*'s pigment enzymes. It takes *E. coli* ten to fifteen hours of exposure to develop an image, which has a rather ghostly appearance. But because each microbe can adjust its own color, the photograph has a very high resolution, about ten times that of a high-resolution printer.

These sorts of experiments give synthetic biologists great hope. Soon it will be possible for them to synthesize entirely new genes from scratch at very little cost. No one can actually invent a completely new gene for a particular function, but it is possible to tinker with existing genes and simulate how their proteins would change as a result. Already researchers have fashioned new genes that allow *E. coli* to detect nerve gas and TNT. One of the most ambitious projects in all of synthetic biology is taking place at the University of California, Berkeley, where scientists have been developing new genetic circuits that may allow *E. coli* or yeast to produce a drug for malaria. The drug, known as artemisinin, is normally produced only by the sweet wormwood plant. If a microbe could make artemisinin, the price might drop by 90 percent.

Meanwhile, Christopher Voigt and his colleagues have created strains of *E. coli* that might someday fight cancer. The microbes seek out tumors by sensing their low levels of oxygen; having found a tumor, they deploy needles to inject toxins into the cancer cells. Voigt hopes someday to turn *E. coli* or some other microbe into a smart drug, able to make its own decisions about when to produce molecules to treat a disorder. Other researchers are trying to turn *E. coli* into a solar battery, able to trap sunlight and turn it into fuel. Synthetic biologists plan to move beyond *E. coli*, just as genetic engineers did. Someday they may be able to hack the programming of human cells, causing them to build new organs.

These are the things synthetic biologists think about when they're in a good mood. When they're in a bad mood, they think about all the challenges they still face.

Engineers, for example, need standardized parts. When engineers design a lathe or a lawn mower, they don't have to design the nuts and bolts that hold the parts together. They just specify which size the nuts and bolts should be. Yet this shortcut is a relatively recent luxury. Before the mid-1800s, the threads on a nut made in one shop might not fit the threads on a bolt made in another. The standardization of those threads sped up the pace of invention and may even have played a major role in driving the Industrial Revolution.

For now, synthetic biology is a craft practiced by artisans. It took Elowitz and his colleagues—some of the world's top experts on *E. coli* and its genes—more than a year to produce blinking bacteria. And once they had their successes, it was very difficult for other scientists to improve their circuits or incorporate them into more elaborate ones. For one thing, a scientist would have to reconstruct the circuit. And the circuit might work only in a particular strain of *E. coli*. Scientists can keep track of *E. coli* strains only with elaborate pedigree charts, tracing the bacteria like royalty. Such are the challenges that make engineers despair.

Since 2001, Drew Endy and Thomas Knight of MIT have been building a catalog of standardized parts for synthetic biology. If you want to add a toggle switch to your particular circuit, you can search for it on the BioBricks Web site, download the DNA sequence, order the corresponding fragments of DNA from a biotech firm, and insert them in *E. coli.* With more than 160 parts in its inventory, BioBricks has not only made synthetic biology easier but has also begun to foster a community. Endy and Knight made BioBricks the basis of the annual synthetic biology competition for students. The students themselves add more parts to the registry, opening the way for future inventions.

But as synthetic biologists try to build more ambitious circuits, they may find a new obstacle in their path: *E. coli* itself. For all of the attention scientists have lavished on it, there is still much about the microbe they do not understand. Six hundred genes remain absolute mysteries. The microbe's genetic network is particularly murky. Scientists can identify most of *E. coli*'s transcription factors, the proteins that grab DNA to switch genes on and off, but they know only about half their targets. And what synthetic biologists do understand about *E. coli* sometimes makes their hearts sink. Its circuits overlap with one another, forming tangles that no self-respecting engineer would ever design. It is very hard to predict how extra circuits will change the behavior of such a messy network.

Some synthetic biologists are trying to overcome *E. coli*'s mystery by taking it apart and rebuilding it from scratch. At Harvard University, for example, George Church and his colleagues have drawn up a list of 151 genes, which they think would be enough to keep an organism alive. Scientists understand these genes—which are drawn mostly from *E. coli* and its viruses—quite well. There should be relatively little mystery when they come together. Church hopes to create a genome with these essential genes. By combining it with a membrane and protein-building ribosomes, he hopes to create a living thing. Call it *E. coli* 2.0.

Meanwhile, at Rockefeller University, Albert Libchaber took an even simpler approach. He and his colleagues cooked up a solution of ribosomes and other molecules found in *E. coli.* Instead of a full genome, they engineered a small plasmid. They then added oily molecules from egg yolks, which form bubbles that scoop up the genes and molecules. These bubbles, Libchaber's team found, could live—at least for a few hours. One of the genes Libchaber added to the plasmids encoded a pore protein. The protocells read the gene, built the proteins, and inserted them in the membrane. There they could allow amino acids and other small molecules to move into the protocell without letting the plasmid and other big molecules out. To track the production of new proteins, the scientists also added a gene from a firefly. The protocells gave off a cool green glow. Libchaber doesn't call his creation a living thing. He prefers the term *bioreactor.* To go from bioreactor to life will take much more work. For one thing, Libchaber and his colleagues will need to add genes to allow the bioreactors to divide into new bioreactors.

Church and Libchaber are only just starting to figure out how to use parts of *E. coli* to create new life-forms. They cannot just throw together DNA and some other molecules and let them come to life on their own. Life is not like a computer, which simply boots up at the press of a button. Every *E. coli* alive today emerged from an ancestor, which emerged from ancestors of its own. Together they form an unbroken river of biology that has flowed continuously for billions of years. Life as we know it has always been part of that river. Perhaps now we will make a canal of our own.

RETURN OF FRANKENSTEIN'S MICROBE

In May 2006, synthetic biologists met in Berkeley, California, for their second international meeting. Along with the standard research talks, they set aside time to draft a code of conduct. The day before, thirty-five organizations—representing, among others, environmentalists, social activists, and biological warfare experts—released an open letter urging that the biologists withdraw the code. They should join a public debate about synthetic biology instead and be ready to submit to government regulations. "Biotech has already ignited worldwide protests, but synthetic biology is like genetic engineering on steroids," said Doreen Stabinsky of Greenpeace International.

These days, biotechnology is experiencing an intense case of déjà vu. The questions people are debating about synthetic biology are strikingly similar to the ones that came up when genetically engineered *E. coli* made news in the 1970s. Do the benefits justify the risks? Is there any intrinsic wrong in tinkering with life? The new debate is far more complex than the old one, in part because *E. coli* is not the only thing scientists are manipulating. Now we must consider transgenic crops, engineered stem cells, human-animal chimeras. The new debate often turns on subtle points of medicine, conservation biology, patent law, and international trade. But for all the differences, the parallels are still powerful and instructive. To understand the potential risks and benefits of the new biotechnology, it helps to look back at the fate of genetically engineered *E. coli* over the past three decades.

The dire warnings that *E. coli* would create tumor plagues and insulin shock epidemics seem quaint today. In thirty years no documented harm from genetically engineered *E. coli* has emerged, despite the fact that many factories breed the bacteria in 40,000-liter fermenters in which every milliliter contains a billion *E. coli.* No one has a God's-eye view of the fate of every engineered *E. coli* in the past thirty years, so it's impossible to know for sure why the predicted plagues never came. Some clues have come from experiments. Scientists put *E. coli* K-12 carrying human genes in tubs of sludge and tanks of water and animal guts. They found that the bacteria rapidly disappeared. Genetically engineered *E. coli* channel a lot of energy and raw materials into making the proteins from inserted genes. But those proteins, such as insulin and blood thinners, probably don't boost *E. coli*'s growth or odds of surviving in the wild. In the carefully controlled conditions scientists create in laboratories, they can thrive. But pitted against other bacteria, they fail.

Genetic engineers did not introduce genes to *E. coli* from other species for the first time. In a sense, *E. coli* and its ancestors have been genetically engineered for billions of years. But most of the transfers have been complete failures. Bacteria cannot make proteins from many horizontally transferred genes, and natural selection favors mutations that strip most alien genes from their genomes.

Unfortunately, the absence of evidence is not a slam-dunk case for the evidence of absence. If an engineered strain of *E. coli* escapes from a factory and manages to survive in the outside world for a few days, it may be able to pass its genes to other bacteria. If a soil microbe picks up a gene for human insulin or some other alien protein, it probably would not benefit from it. But the possibility can't be ruled out. Studies suggest that even if an alien gene gave bacteria a competitive advantage, it would remain too rare for scientists to detect for decades, perhaps even centuries.

While we've been waiting for a genetically engineered monster to emerge, *E. coli* O157:H7 has emerged as a serious threat to public health. It was in 1975—the same year in which scientists gathered at Asilomar to ponder the potential dangers of genetically engineered *E. coli*—that a woman suffered the earliest known attack of *E. coli* O157:H7. But that pathogen was not the work of a human genetic engineer with an intelligent design. Over the course of centuries, *E. coli* O157:H7 acquired many genes from viruses carrying deadly instructions. They acquired these genes from other strains of *E. coli* or other species of bacteria. They acquired syringes and toxins and molecules that alter the behavior of host cells. This genetic engineering is still taking place as one new strain after another evolves. But the insertion of a bundle of genes in a single microbe was only the first step in this transformation. Natural selection then had to favor those genes in their new host; it had to fine-tune them.

The transformation required an entire ecosystem that could produce the conditions that would drive natural selection. We provided it. *E. coli* O157:H7 had been pumped from humans to livestock through farm fields and slaughterhouses, through rivers and sewers rife with toxin-bearing viruses. There's little evidence for a similar evolutionary pump for genetically engineered *E. coli*. Our unplanned engineering of *E. coli* may give us more to worry about than anything brewed up in a lab.

Thirty years have passed since the backers of genetic engineering predicted recombinant DNA would bring great rewards. They were right, up to a point. *E. coli* and other engineered cells not only produce a vast number of valuable molecules; they have also sped up the pace of science enormously. *E. coli* was a crucial partner in the sequencing of the human genome, for example. In order to read the genome, scientists inserted chunks of it into *E. coli*, which then produced many copies that scientists could analyze. Other scientists have used *E. coli* to churn out millions of proteins so that they can discover what the proteins do. By inserting human genes into *E. coli*, scientists discovered that they are made up of two kinds of DNA. Some segments of the genes, known as exons, encode parts of proteins. But they alternate with other segments, called introns, that encode nothing. Our cells edit out the introns from RNA in order to make proteins. They can even use different combinations of exons to produce a number of proteins from a single gene.

As important as these accomplishments have been, however, genetic engineering has fallen far short of the more extravagant promises offered thirty years ago. Cetus predicted that all major diseases would surrender to genetically engineered proteins by 2000. I'm writing in 2007, and cancer, heart disease, malaria, and other diseases continue to kill by the millions. Maybe the people at Cetus were just wrong about the date. Perhaps another thirty years will bring some major breakthrough in genetic engineering that will wipe out all major diseases. I wouldn't bet on it, though. Most major diseases are fiendishly complex, and a single engineered protein is not going to make them go away. Diabetes, the poster child for the promise of genetic engineering, has not disappeared over the past thirty years. In fact, it has exploded. The incidence of type 2 diabetes has doubled in the United States, and cases of diabetes worldwide have increased tenfold. *E. coli* has provided insulin for millions of people with diabetes, but, as Ruth Hubbard warned, it did nothing to prevent the disease. Genetic engineering could not block the sources of the diabetes epidemic, which may include the availability of cheap sugar. That sugar comes increasingly from high-fructose corn syrup, whose low price we owe to breakthroughs in genetic engineering.

Drugs made through genetic engineering have also turned out to be just as vulnerable to market forces as conventional ones. Drug companies have been trying to increase their sales by expanding our definition of what it means to be sick. Genetically engineered drugs have been promoted this way as well. Genentech originally got approval from the Food and Drug Administration to sell its *E. coli*–produced growth hormone to treat children whose bodies couldn't make it themselves. But in 1999 the company had to pay $50 million to settle charges that its drug was being marketed to children who were merely shorter than average.

E. coli's thirty-year history of genetic engineering is worth considering when we judge the new biotechnology that has come in its wake. We must resist empty fear and empty hype. We must instead be realistic, always remembering how both nature and society actually work.

One of the great dreams of biotechnology has been to end famine, for example. Julian Huxley speculated as far back as 1923 that scientists would create a limitless supply of food (along with purple oceans). The dream lived on in the 1960s with promises of oil-fed yeast. When scientists successfully inserted foreign genes in *E. coli,* advocates for genetic engineering promised more food for a starving world. In the 1970s, the Green Revolution—the result of breeding new varieties of crops and using plenty of fertilizer—had dramatically increased farm productivity. But the world's population, and thus its hunger, were still growing. Scientists began trying to engineer bacteria to make fertilizer by capturing nitrogen from the air. Most recently, scientists have turned their attention to engineering plants themselves. Transgenic crops are being promoted not as a way to make bigger profits but as a way to fight hunger and malnutrition. Crops that can resist viruses and insects will increase harvests. Crops that can resist herbicides will allow farmers to fight weeds more effectively, increasing the yield even more. Norman Borlaug, who won a Nobel Peace Prize for his work on the Green Revolution, claimed that genetically modified crops would pick up where his own work had left off, feeding the world for another century.

Anyone who questioned this prediction, Borlaug suggested, was dooming the world's poor to famine. "The affluent nations can afford to adopt elitist positions and pay more for food produced by the so-called natural methods; the 1 billion chronically poor and hungry people of this world cannot," he wrote in 2000. "New technology will be their salvation, freeing them from obsolete, low-yielding, and more costly production technology."

One of the promising crops Borlaug—as well as many other advocates—pointed to was Golden Rice, a strain of rice engineered to make vitamin A. Vitamin A deficiency affects roughly 200 million people worldwide. Up to half a million children become blind each year, half of whom will die within a year of losing their sight. In the late 1990s, Swiss scientists began inserting genes from daffodils and bacteria into the rice genome to produce vitamin A. They formed a partnership with the corporation Syngenta to develop the rice and distribute it free to farmers who make less than $10,000 a year. Ingo Potrykus, one of the inventors, appeared on the cover of *Time* in 2000, alongside the headline "THIS RICE COULD SAVE A MILLION KIDS A YEAR," which was followed in small print by ". . . but protesters believe such genetically modified foods are bad for us and our planet. Here's why."

Potrykus had little patience for those protesters. "In fighting against 'Golden Rice' reaching the poor in developing countries," he declared in 2001, "GMO opposition has to be held responsible for the foreseeable unnecessary death and blindness of millions of poor every year."

Strong words, particularly given how embryonic the research on

Golden Rice was when Potrykus uttered them. He and his colleagues had published their first results only the previous year. They had managed to produce only small amounts of vitamin A in the rice's tissues, far too little to wipe out vitamin A deficiency. In 2005, four years after Potrykus accused his critics of mass murder, Syngenta scientists discovered that adding an extra gene from corn helped boost the level of the vitamin A precursor more than twentyfold. It was a huge increase, but there's no solid evidence yet of how much benefit it brings to people who eat it. Some nutritionists have warned that it may not bring much benefit at all, because vitamin A has to be consumed along with dietary fat in order to be properly absorbed by the body. It's possible to suffer vitamin A deficiency—even to go blind—on a diet that contains vitamin A. Foods such as milk, eggs, and many vegetables offer the right combination of vitamin A and fat, but rice does not. Just because Golden Rice is at the cutting edge of genetic engineering doesn't mean that it will cut down vitamin A deficiency any more than conventional methods have.

Using words like *salvation* to describe transgenic crops makes as little sense as calling them Frankenfoods. We are thrown back and forth between the extremes of abject terror and hope for miracles of loaves and transgenic fish. Genetically modified crops are hardly miraculous. They are living things, as much subject to the rules of life as *E. coli* or humans. And just as *E. coli* has evolved defenses against some of our best antibiotics, natural selection is undermining the worth of the most popular transgenic crops.

About 80 percent of all the transgenic crops planted in 2006 were engineered for the same purpose: to be resistant to a herbicide known as glyphosate. Glyphosate kills plants by blocking the construction of amino acids that are essential to their survival. It attacks enzymes that only plants use, with the result that it's harmless to people, insects, and other animals. And unlike other herbicides that wind up in groundwater, glyphosate stays where it's sprayed, degrading within weeks. A scientist at the Monsanto Company discovered glyphosate in 1970, and the company began selling it as Roundup in 1974. In 1986, scientists engineered glyphosate-resistant plants by inserting genes from bacteria that could produce amino acids even after a plant was sprayed with herbicides. In the 1990s, Monsanto and other companies began to sell glyphosate-resistant corn, cotton, sugar beets, and many other crops. Instead of applying a lot of different herbicides, farmers found they could hit their fields with a modest dose of glyphosate alone, which wiped out weeds without harming their crops. Studies indicated that farmers who grew the transgenic crops used fewer herbicides than those who grew nontransgenic plants—77 percent fewer in Mexico, for example—while getting a significantly higher yield.

For a while it seemed as if glyphosate would avoid the fate of many other herbicides before it: the evolution of weeds resistant to herbicides. Glyphosate seemed to strike at such an essential part of their biology that no defense could possibly evolve. Of course, it also seemed for a while as if *E. coli* couldn't evolve resistance to Michael Zasloff's antimicrobial peptides. And after glyphosate-resistant crops had a few years to grow, farmers began to notice horseweed and morning glory and other weeds encroaching once more on their fields. Farmers in Georgia have had to destroy fields of cotton because of infestations of resistant Palmer amaranth. When scientists have studied these resurgent weeds, they've discovered genes that now make the plants resistant to glyphosate.

There's no evidence that these weeds acquired their resistance from the transgenic crops. They most likely got it the old-fashioned way: they evolved it. Using glyphosate on transgenic crops proved to be so cheap and effective that farmers flooded huge swaths of land with a single herbicide. They created an enormous opportunity for weeds that could resist glyphosate and drove the quick evolution of stronger and stronger resistance. And once the weeds evolved their resistance, they appear to have passed on the resistance genes to other weedy species.

When antibiotics fail against *E. coli* and other bacteria, it may take years for a new kind of antibiotic to emerge. The pipeline of transgenic crops is equally sludgy. It wasn't until 2007, more than twenty years after the invention of glyphosate-resistant crops, that scientists announced they had engineered plants with genes that make them resistant to another herbicide, known as dicamba. Monsanto licensed the technology but said it wouldn't have dicamba-resistant crops ready for sale for another three to seven years. In the meantime, farmers can resort to old-fashioned methods to slow the evolution of resistance, rotating crops and using a combination of herbicides.

Although there's a lot of déjà vu in biotechnology today, some scientists have been carefully studying the fate of *E. coli* in the 1970s in order to avoid some of the mistakes their predecessors made. Synthetic biologists have become particularly keen historians, learning how the pioneers in their field grappled with risks, regulations, and the public perception of their work. Rather than make synthetic biology the privileged domain of an elite, Drew Endy and his colleagues are inviting the public to join in the experience. Anyone can download the codes for BioBricks. The *E. coli* camera is now appearing in science museums, and high school students are entering synthetic biology competitions. And rather than put all their efforts into creating a big moneymaker like insulin, synthetic biologists are trying to make cheap drugs for malaria, to demonstrate the good that can come of their work.

Synthetic biologists want to preserve this open-source spirit despite the fact that their tools may someday be used for evil ends. It's conceivable, for example, that a government might design an organism for biological warfare. Synthetic biologists fear that if the government takes over their research, innovations will dry up. They argue that the best way to defeat an engineered pathogen is to harness the collective creativity of an open community. By keeping synthetic biology free of excessive regulations and patents, its founders hope they can foster an artificial version of the open-source evolution that has served *E. coli* so well for millions of years.

"IT IS CONFUSION"

In the 1970s, genetically engineered *E. coli* frightened people not just with its potential risks. It touched something deeper—a feeling that genetic engineering is something humans were simply not meant to do. Genetic engineering would disrupt the order of nature, the result of billions of years of evolution. Shuttling genes or other biological material from species to species would blur barriers that had been established long before humans existed, threatening to tear down the very tree of life.

"We can now transform that evolutionary tree into a network," declared Robert Sinsheimer, a biologist at the University of California, Santa Cruz. "We can merge genes of most diverse origin—from plant or insect, from fungus or man as we wish." Humans, Sinsheimer believed, were not prepared for this responsibility: "We are becoming creators—

makers of new forms of life—creations that we cannot undo, that will live on long after us, that will evolve according to their own destiny. What are the responsibilities of creators—for our creations and for all the living world into which we bring our inventions?"

One newspaper called genetic engineering on *E. coli* "the Frankenstein project." Tampering with DNA, the MIT biologist Jonathan King declared, was "sacrilegious." Two political activists, Ted Howard and Jeremy Rifkin, condemned genetic engineering in a 1977 book called *Who Should Play God?*

Thirty years later, critics of biotechnology continue to play the Prometheus card. In 1999, for example, Rifkin organized a full-page ad representing a number of organizations that were demanding controls on biotechnology. The ad, which appeared in *The New York Times*, displayed two examples of the new horrors humanity faced: a human ear growing from the back of a mouse and the first cloned animal, a sheep named Dolly. Across the top of the ad was the headline "Who Plays God in the Twenty-first Century?"

> The genetic structures of living beings are the last of Nature's creations to be invaded and altered for commerce. . . . Does anyone think it's shocking [that the] infant biotechnology industry feels it's okay to capture the evolutionary process, and to reshape life on earth to suit its balance sheets? . . . to take over Nature's work? . . . Whether you give credit to God, or to Nature, there is a boundary between life forms that gives each its integrity and identity.

"To God, or to Nature"—an intriguing choice. It is certainly true that Christianity and Judaism have an uneasy relationship with biotechnology. After all, in the first pages of Genesis, the Bible makes the essences of species paramount:

> And God said, let the earth bring forth grass, the herb yielding seed, and the fruit tree yielding fruit after his kind. . . . And God created great whales and every living creature that moveth, which the waters brought forth abundantly after their kind, and every winged fowl after his kind. . . . And God said, let the earth bring forth the living creature after his kind, cattle and creeping thing, and beast of the earth after his kind, and it was so.

In Leviticus, humankind is instructed to keep those distinctions clear: "Thou shalt not let thy cattle gender with a diverse kind: thou shalt not sow thy field with mingled seed."

The one kind of life most important of all in the Bible is, of course, our own. Made in God's image, we must never come close to blurring the distinction between us and animals: "Neither shalt thou lie with any beast to defile thyself therewith: neither shall any woman stand before a beast to lie down thereto: it is confusion."

For many conservatives today, biotechnology's threat to human nature, rather than to nature, is most alarming. "Using human procreation to fuse animal-human runs counter to the sacredness of human life and man created in the image of God," writes Nancy L. Jones of the conservative Center for Bioethics and Human Dignity.

Some conservatives don't cite chapter and verse, but they agree that crossing the species barrier degrades human nature. The most prominent of these critics is Leon Kass. After his encounter with Paul Berg in the early 1970s, Kass continued to write and speak about bioethics, and from 2002 to 2005 he was the chairman of President George W. Bush's Council on Bioethics. In his arguments against chimeras and cloning, he says that the gut feeling that there's something disgusting about them is its own evidence that they're wrong. Kass calls this reliable disgust "the wisdom of repugnance." We just *know* that certain things are wrong, such as incest and mutilating a corpse. Our inability to give a rational explanation for our feelings does not deny their importance.

In fact, Kass argues, this disgust is a valuable guide to what we should embrace and reject. There's something horrifying about an army of human clones or human-animal chimeras. In an age when technology can provide us with so much, Kass has written, "repugnance may be the only voice left that speaks up to defend the central core of our humanity. Shallow are the souls that have forgotten how to shudder."

Theologians and philosophers are not the only people making these sorts of arguments. In January 2006, President Bush called for a ban on "animal-human hybrids," adding that "human life is a gift from our creator, and that gift should never be discarded, devalued or put up for sale." A bill to ban chimeras, introduced by Senator Sam Brownback of Kansas, states that "respect for human dignity and the integrity of the human species may be threatened by chimeras."

To tamper with the essence of human nature—by introducing human brain cells into a mouse, for example, or by altering the genes in a fertilized egg—would be to degrade what it means to be human. In the words of Robert George, a Princeton political scientist and a member of Bush's Council on Bioethics, "A thing either is or is not a whole human being."

To make sense of these arguments, it helps to look back once again at *E. coli.* Thirty years ago, engineering *E. coli* was considered an affront to nature, even to God. It defied billions of years of evolution by sporting a gene from a human. Now no one seems to care about it. *E. coli* sits neglected in its fermenting tanks and laboratory flasks, loaded with imported genes from hundreds of other species, including our own. *E. coli* starves and suffers as it churns out alien proteins. And yet it no longer offends the wisdom of our repugnance. There are no campaigns to respect the integrity of *E. coli* as a species, to fight the degradation of human nature that comes from putting human genes into bacteria. It's hard to imagine someone turning down a prescription for blood thinner because it is the product of the unholy union of human and microbe.

How can our fear of crossing species boundaries be so strong and yet so mutable? It does not arise from an objective perception of some deep, incontrovertible fact of life. It is a habit of mind. We are all intuitive biologists from childhood. Babies quickly come to expect differences between living things and nonliving ones. Rocks tumble under the force of gravity, for example, but an ant crawls by its own agency. As children grow, they come to recognize different kinds of living things—animals and plants, for example, or cats and dogs. Each kind has its own essence, an invisible force that produces its actions. This intuitive biology comes easily to children, without elaborate training. And it becomes the habitual way in which adults think about life.

Intuitive biology may have evolved as an adaptation of the human mind, like language and color vision. It may have helped our ancestors organize their understanding of the natural world. The more knowledge our ancestors could gain about animals and plants, the more likely they were to find food and survive. They could predict where to find wildebeests at a certain time of the year, when to look for tubers in the ground, which kinds of fruit were poisonous and which were sweet. Our ancestors became keen connoisseurs of subtle differences between species, such as colors and coat patterns. Those differences could mean the difference between life and death, between eating a poisonous berry and escaping starvation.

The notion of the integrity of species emerges from our intuitive biology. Even to dream of breaking the species barrier can stir up strong emotions. It's striking that some of the earliest artwork made by our species includes chimeras. Some 30,000 years ago, for example, a sculptor in Germany carved a piece of ivory into the form of a lion-headed woman. The image, perhaps seen in a dream or a trance, must have had a profound mystical meaning to the sculptor and to all who looked at it. It blurred the essences of species. By violating the rules of intuitive biology, it became magical. Magical hybrids—including the original Chimera, a monster from Greek mythology, part goat, part lion, part snake—turn up again and again throughout history.

Now modern biology has challenged our intuitive biology. Species are no longer immutable essences but the products of evolution. Darwin argued that humans descended from apes, which descended from older mammals, all the way back to blind, jawless fish. For breaking the rules of intuitive biology, Darwin was punished by being turned into a chimera. Cartoonists drew him with the bearded head of a man and the hairy body of a monkey.

In 1896, H. G. Wells played on this anxiety with his novel *The Island of Dr. Moreau.* Dr. Moreau, his sense of morality lost in the lust for scientific knowledge, surgically combines different animals into humanlike monsters.

"The thing is an abomination," the narrator declares to Moreau. The evil doctor replies, "To this day I have never troubled about the ethics of the matter. . . . The study of Nature makes a man at last as remorseless as Nature."

Wells punishes Moreau for his transgression with an uprising of the monsters. *The Island of Dr. Moreau* is a prophetic book, especially given that Wells wrote it before biologists had discovered genes. Once scientists understood DNA, it became the new essence of life. Today our true selves lie in our genes. The origin of our genome at conception becomes the origin of a new life. DNA has also come to define the essence of a species, what distinguishes it from other living kinds. Thus came a horror at the thought of mingling genes from different species, particularly species that look as different from each other as humans and *E. coli.* Genetic engineering defies a powerful rule we use to organize the living world. Setting the boundaries of species is not the business of humans. When humans tamper with those boundaries, they create monsters, they unleash horrors.

Our intuitive biology did not evolve because it was true. It evolved because it was useful. It allowed our ancestors to make good decisions based on the information they could gather, and those decisions raised their odds of surviving and reproducing. But intuitive biology is not a reliable guide to the deep truths of life. What is the essence of *E. coli* as a species, for example? It's not being a harmless, sugar-feeding, flagella-producing microbe. Within the species we call *E. coli,* you can also find aggressive defenders of the gut that shut out disease-causing pathogens. You can find many pathogens equipped with weapons not found in harmless strains. Some strains straddle the divide—they are beneficial, but they also carry many of the genes that make other strains killers. And many of these strains evolved by being infected with viruses that show no respect for our beloved species boundaries. There is no immutable essence that unites *E. coli.*

Our intuitive biology fails us when we try to understand *E. coli,* and it also fails us when we try to understand ourselves. Like all other living species, humans are the product of evolution. If we weren't, the entire controversy over biotechnology would not exist in the first place. If human nature were truly distinct, it would be impossible to plug human genes so easily into *E. coli* or to grow human brain cells in a mouse's skull. The essence of being human is as much a construction of our minds as the essence of *E. coli.*

New research on human evolution makes it impossible to believe that a thing either is or is not a whole human being, as Robert George has claimed. Consider a gene called microcephalin. There are several versions of the gene floating around our species, but one is far more common than the others, found in 70 percent of all people on Earth. Scientists at the University of Chicago decided to trace the history of this version of microcephalin. They found compelling evidence that it entered the human genome long after *Homo sapiens* had evolved.

About half a million years ago, our ancestors split off from the ancestors of Neanderthals. The split probably occurred in Africa. Afterward, the ancestors of Neanderthals spread across Europe, while the forerunners of our species stayed behind in Africa. *Homo sapiens* evolved about 200,000 years ago. It was only after our species emerged that humans evolved full-blown language, abstract thought, the capacity for art, and many of the other qualities that are at the core of what we call human nature.

About 40,000 years ago, *Homo sapiens* expanded their range into Europe. And there humans encountered Neanderthals. Neanderthals became extinct about 28,000 years ago, but it appears that before they disappeared they interbred with humans. Most of their genes disappeared over the generations, but at least one survived: their version of microcephalin. It didn't just survive, in fact—it spread like wildfire. Something about it was strongly favored by natural selection, with the result that it now can be found in the majority of humans alive today. And microcephalin isn't some minor gene for growing nose hair or coloring toenails. It plays a central role in the development of the brain. Thanks to natural engineering, most humans carry this nonhuman gene, which is involved in building that most human of organs, the brain. By George's reasoning, most humans are not human.

Hybridization is not the only way foreign DNA got into our cells. Some 3 billion years ago our single-celled ancestors engulfed oxygen-breathing bacteria, which became the mitochondria on which we depend. And, like *E. coli,* our genomes have taken in virus upon virus. Scientists have identified more than 98,000 viruses in the human genome, along with the mutant vestiges of 150,000 others. Some have donated their DNA to our own biology, such as the placenta. If we were to strip out all our transgenic DNA, we would become extinct. Some of these viruses inserted copies of themselves after our split from chimpanzees. Some are found in Asians and Europeans but not in Africans, suggesting that they infected the human genome only after some humans emerged from Africa 50,000 years ago. When people acquired this foreign DNA, did they lose their human nature?

It is awkward to think this way. It feels unnatural. The unnaturalness is in the workings of our minds, however, not in nature. But we will probably get used to it, in the same way we have gotten used to thinking of matter as being made up of subatomic particles. Our repugnance toward breaches in the species barrier and toward the modification of genes is shifting even now. The lack of angry mobs trying to burn down insulin-producing factories to preserve the natural order of things is proof of that.

This sort of change may well disturb a critic like Leon Kass. In 1997, he testified before Congress in favor of a ban on human cloning, declaring, "In a world whose once-given natural boundaries are blurred by technological change and whose moral boundaries are seemingly up for grabs, it is, I believe, much more difficult than it once was to make persuasive the still compelling case against human cloning. As Raskolnikov put it, 'Man gets used to everything, the beast!' "

There's a contradiction here. On the one hand, our wisdom of repugnance is supposed to be a deeply anchored, reliable guide to what is fundamentally right and wrong—not what happens to be right and wrong this afternoon. On the other hand, Kass is angry that this sort of repugnance can disappear as times change. It's hard to see how he can have it both ways.

We can be overwhelmed by our emotional reactions to scientific advances. In some cases, we eventually recognize that we were probably right—or wrong—to have those feelings. In other cases, our perception of essences triggers feelings of disgust when those essences seem to be corrupted. That disgust may be triggered by *E. coli* carrying human genes, or in vitro fertilization, or a person receiving a heart valve from a pig. But as we come to recognize the benefits or risks of those developments, as we see the world not coming to a Pandora's-box end, our sense of disgust fades.

We don't become Dostoyevskian beasts along the way, though. With the advent of organ transplants, we did not slide down a slippery slope into a world in which paraplegics have their livers yanked out against their will. There are certainly new choices to make—to allow the sale of organs or not, for example—but we continue to make them seriously.

Chimeras and various sorts of genetic engineering will become more common, but they will not, I suspect, produce a moral meltdown. For one thing, a lot of the most startling nightmare scenarios we hear about today have little basis in science. Mice with human neurons will not cry out, "Help me, help me!" There is much more to being human than possessing a peanut-sized clump of neurons. Yet we may decide that engineering such a mouse is cruel to the animal itself. (Repugnance at cruelty toward animals is actually a new sort of disgust many people have acquired, rather than lost, over the past 200 years.) And some chimeras will probably be banned because the challenges they pose to our moral treatment of humans and animals don't justify the procedure.

I suspect—or at least I hope—that as we make these decisions, we will come to a deeper understanding of what it means to be human: not as an inviolable essence but as a complex cloud of genes, traits, environmental influences, and cultural forces. If we do gain this wisdom, it may turn out to be the most important gift *E. coli* has given us.

Eleven

N EQUALS 1

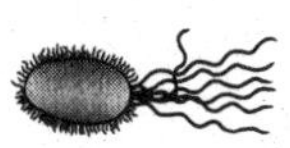

I AM STANDING IN MY YARD on a winter night, looking up at a few bright stars asserting themselves against a gibbous moon. I hold up a petri dish of *E. coli* against the sky. The moonlight shines through the leafless maples into the agar. It gives the colonies a cool, cloudy glow. They look like worlds and stars. I have reached the final question about *E. coli,* a twist on Monod's old boast. Is everything that is true for *E. coli* true for an alien?

One night in October 1957, Joshua Lederberg looked up at the stars as well. He was in Australia, where he was spending a sabbatical. Lederberg was only thirty-two at the time, but he had more than a decade of research behind him, for which he would win a Nobel Prize the following year. He had done most of that work on *E. coli.* He had discovered that the microbe had sex, and he had used its sex life to draw some of the first maps of its genes. He and his wife had confirmed that genes mutate spontaneously, helping to bring Darwin into the molecular age. They had discovered viruses that could merge into their *E. coli* hosts. Thanks in large part to Lederberg, *E. coli* was becoming the standard tool for studying the molecular basis of life, and other scientists were beginning to use it to translate the genetic code.

Now Lederberg was restless. He had come to Australia, to the University of Melbourne, to study the immune system. White blood cells learn to recognize bacteria and other parasites, but they don't use ordinary genes to encode those lessons. No one at the time knew what language they used. Lederberg would return to the United States recharged, but white blood cells would not be his obsession. Instead, it would be space.

On that night in the Australian spring, Lederberg had gazed up at a moving point of light. It was not a star or even a meteorite but a steel ball hurled into space by humans. Lederberg had a hunch that the Soviet Union's launch of the first Sputnik satellite was going to change the world.

Lederberg saw in space travel a new frontier for molecular biology. He and other molecular biologists were in the midst of discovering just how uniform life on Earth actually is. *E. coli* and elephants both encode genes with DNA, both use RNA to carry that information to ribosomes, and both use the same genetic code to translate it into proteins. The uniformity of life was a staggering discovery, Lederberg later wrote, "but its domain has been limited to the thin shell of our own planet, to the way in which one spark of life has illuminated one speck in the cosmos." Only by going to other worlds would scientists be able to learn whether a similar kind of life had emerged beyond Earth.

Lederberg worried that this awesome opportunity would be ruined if the United States and the Soviet Union ended up in a heedless race into space. In their rush to plant a flag on the moon or Mars, they might contaminate other worlds with microbes from Earth. When Lederberg returned to the United States, he began to lobby the newly formed National Aeronautics and Space Administration to treat outer space like a petri dish, to be kept free of contamination.

He quickly organized meetings at which scientists debated the potential risks of space travel. Unless special precautions were taken, they agreed, a visit to another planet would inevitably leave bacteria there. An astronaut would be "a teeming reservoir of microbial contamination," as Lederberg wrote. Unmanned probes might pick up millions of bacteria from their human engineers, which they could carry to another world.

A 1959 panel of scientists tried to imagine what would happen if a single *E. coli* arrived on a planet devoid of life but rich in organic carbon. "The common bacterium *Escherichia coli* has a mass of 10^{-12} grams and a minimum fission interval of 30 minutes," they wrote. "At this rate it would take 66 hours for the progeny of one bacterium to reach the mass of the Earth. The example illustrates that a biological explosion could completely destroy the remains of prebiological synthesis."

Lederberg's efforts ultimately led to an agreement between the United States and the Soviet Union on standards for sterilizing spacecraft. Yet Lederberg became famous not for his worries about contaminating other planets but for his worries about the return trip. If life did exist on other worlds, a spacecraft coming back to Earth might accidentally carry some of it home. Alien microbes might wreak havoc on our planet. They might cause a global plague or trigger a famine by attacking crops.

"The fate of mankind could be at stake," Lederberg warned. Soon reporters were describing the dire warnings of the Nobel Prize–winning biologist, using headlines such as "Invasion from Mars? Microbes!" A version of Lederberg even ended up in the 1971 science-fiction movie *The Andromeda Strain:* the intrepid biologist desperately trying to find a cure for a virus from outer space.

For all his worries, however, Lederberg did not want to seal off the sky. At NASA's invitation, he set up a laboratory at Stanford University to begin building a device that could detect signs of life on another planet. In some ways the work was mundane. Lederberg and his colleagues tinkered with conveyor belts and mass spectrometers. But they also faced a profound question, less scientific than philosophical: How can you search for life you've never seen? The question, Lederberg decided, required a new branch of biology all its own. He dubbed it exobiology, the biology of life beyond Earth.

The goal of exobiology was to discover whether life has begun more than once in the universe and whether it has taken more than one form. Does life *have* to use DNA? Does it *have* to build its cells from protein? Is there something about these molecules that suits them to life, something no other combination of atoms can possibly have? "These questions might be answered in two ways," Lederberg wrote. "Presumptuous man might mimic primitive life by imitating Nature, furnishing substitute compounds. More humbly, he might ask Nature the outcome of its own experiments at life, as they might be manifest on other globes in the solar system."

Looking for unearthly forms of life would be difficult because scientists could not predict what they might find. Lederberg felt content starting off with a more conventional search. "We can defer our concern for such exotic biological systems until we have got full value from our searches for the more familiar," he wrote.

NASA agreed. The agency would search for the familiar, and it would search for it on Mars. Mars was just enough like Earth to offer some hope of harboring life. In 1965, *Mariner 4* became the first probe to send back detailed pictures of the surface of Mars. It revealed a bleak landscape, pocked with craters and devoid of forests and other signs of life. If life did exist on Mars, it probably just consisted of microbes. NASA used the pictures from *Mariner 4* and later probes to design a mission to land a probe on the surface of Mars. On July 20, 1976, nineteen years after Lederberg watched the first satellite rise from Earth, *Viking 1* became the first probe to land on another planet.

Sadly, the mission was generally agreed to be a bust. *Viking 1* found no signs of organisms that could convert carbon dioxide to organic carbon. Some kinds of terrestrial life, such as *E. coli,* consume organic carbon and release carbon dioxide as waste, but *Viking* found no trace of this metabolism either. One last experiment remained, a final court of appeals. *Viking* scooped up soil, heated it up to liberate molecules, and then fired them down a tube, where they could be measured. The probe could detect no organic carbon in the Martian soil whatsoever. This result was devastating, because life has created huge amounts of organic carbon on Earth, not just in the bodies of living things but in the waste they leave behind.

"That's the ball game," said Gerald Soffen, the Viking project scientist. "No organics on Mars. No life on Mars."

It appeared that the surface of Mars was far harsher than scientists had reckoned. Ultraviolet light and highly reactive chemicals such as hydrogen peroxide quickly destroyed any organic carbon. The chances of life on Mars seemed low or nil. Lederberg was more optimistic than some of his colleagues, but not by much. It was possible that life on Mars existed only in a few oases, perhaps around hot springs bubbling up from the interior of the planet. But if there was life on Mars, it was far more retiring than the boisterous, all-consuming life on Earth. "We can no longer be confident that no matter where you look you will find life," Lederberg told reporters.

Viking's failure was no reason to stop looking for life, Lederberg and others believed. They urged NASA to put together a "son of *Viking*"—a new probe that could take a new set of instruments to Mars. But NASA was more interested in astronauts, those teeming reservoirs of *E. coli.* As support for exobiology faded, Lederberg returned to other pressing issues in biology, such as the emergence of new diseases and the threat of biological warfare. His days of professional stargazing were over.

Twenty years later, NASA's interest in extraterrestrial life grew again. A meteorite from Mars bore strange markings that some scientists suggested were fossils of microbes. The *Galileo* probe passed by Europa, a moon of Jupiter, and captured images of the ice covering its surface. Perhaps life was lurking underneath. The search for life—now called astrobiology—found new support from NASA, which founded the NASA Astrobiology Institute in 1998.

Today many astrobiologists search for extreme places on Earth where life manages to survive. *E. coli* is a rugged creature, but scientists have found many other organisms that live in places where it would quickly die: acid-drenched mine shafts, oxygen-free swamp bottoms, the depths of glaciers, superheated water shooting out of hydrothermal vents, the spaces inside crystals of salt. Planets and moons with similar environments might be suitable homes for life.

But as weird as some new species may be, they all share *E. coli*'s fundamental features. They are membranes wrapped around proteins and DNA. They need sources of carbon and energy in order to grow. And they need liquid water as a medium in which their chemistry can take place. If some of these rugged microbes were carried to an underground hydrothermal system on Mars or perhaps slipped beneath the icy crust of Europa, they might be able to eke out an existence.

Yet scientists are also keenly aware that life on Earth may not be the rule for life in the universe. Our own tinkering with life has made that clear. Expanding *E. coli*'s genetic code does not kill it, so there's no reason to think that life on other planets couldn't use other amino acids to build its proteins. All life on Earth uses the four-letter language of bases to encode information in its genes. But scientists have been able to engineer *E. coli* with man-made bases—in other words, adding new letters to its alphabet. Synthetic biology blurs into astrobiology.

Life might not even need DNA. Some experiments have suggested that other molecules can take on the same structure, with a backbone carrying information-bearing compounds. They might even be able to replicate themselves accurately. Scientists have even speculated that life may be able to exist without liquid water. Another liquid, such as liquid methane, might serve as its matrix.

No matter what extraterrestrial life might be made of, our discovery of it would change how we think about life in general. It would finally give scientists more than one planet's worth of life with which to search for the rules of existence. Scientists would probably start studying alien life at its lowest levels, trying to determine how it stores genetic information. But some of the most interesting comparisons would come later. Living things on Earth have more in common than DNA. *E. coli* and elephants alike can survive in a changing world thanks to the robust wiring of their genetic circuits. Natural selection shapes their life spans and drives their complex social life, filled with sacrifice and deception. Barriers slice life up into individual organisms, but viruses weave them together in a genetic matrix. Alien life would let us see just how universal these features are.

If alien life were to prove Earth-like, scientists would be faced with two possibilities: Perhaps the same biology emerged independently on different worlds. Or perhaps it went from one world to another.

Anaxagoras, a Greek philosopher who lived in the fifth century B.C., declared that all life on Earth originated from seeds that pervaded the cosmos. He called the process panspermia. In the twentieth century, Francis Crick and several other prominent scientists revived panspermia in various forms. They suggested that spores had fallen to Earth billions of years ago and given rise to all life. The panspermians met with skepticism because they had no clear evidence that life existed on other planets or that it could survive an interplanetary journey. Panspermia was unsatisfying as a theory, because it did not explain the origin of life. It just pushed the question back.

Panspermia still meets with skepticism, but scientists now regularly talk about it at conferences without being laughed off the dais. Early in the history of the solar system, large meteorites were crashing into planets quite frequently, launching material out into space. In some cases, that material could have reached other planets. The path from Mars to Earth is particularly easy because the planets are so close to each other and because Mars has a much weaker gravitational field. Even today an estimated fifteen meteorites from Mars land on Earth each year. Planets may trade bits of themselves over far greater distances. A few Earth rocks could travel all the way to the moons of Saturn and Jupiter. In fact, according to one estimate, a rock from Earth might strike Jupiter's moon Europa once every 50,000 years. To us 50,000 years may be an unimaginably long time, but in the history of the solar system it's like the patters of a hailstorm.

If these studies are correct, it's possible that some *E. coli* rode a meteorite into space thousands of years ago. For most microbes this sort of journey would be fatal. Many would be destroyed by the harsh interplanetary radiation from which our atmosphere shields us. Still others would die in their blazing descent to another world. But a few microbes might survive. And as Lederberg and his colleagues recognized, it would take only a few microbes to populate a fertile planet. Some scientists have even suggested that these journeys might have kept life from disappearing from the solar system altogether. A big enough impact could boil off the oceans of Earth, leaving it sterilized. It would take millions of years for the water vapor to rain back down and allow a stable habitat to form. Life could hold out during that time on Mars or in some other refuge.

The most extreme form of panspermia was proposed in 2004 by William Napier, an Irish astronomer. He argued that some rocks lofted from our solar system might fly out of the solar system altogether. Once safely distant from the sun, the microbes they carried would no longer be harassed by ultraviolet radiation. Some of the rocks might wind up on planets orbiting other stars, and a few of the microbes might find a new place to grow. Of course, those planets would be hit by heavenly bodies as well, and their organisms would be passed on to other solar systems. Napier estimates that this interstellar infection could contaminate the entire galaxy in a few billion years.

Which brings me back to the dish of *E. coli* I hold up to the sky. On some nights at some places on Earth you can spot the International Space Station through a telescope. *E. coli* is up there. It floats inside the bodies of the astronauts, swims in their drinking water, and drifts inside droplets that cling to the space station walls. Has *E. coli* gotten any farther? Lederberg's worry about contaminating other planets has not gone away. No matter what measures engineers take as they build unmanned probes, it seems that a few hardy species manage to settle on their surfaces.

"The field is haunted by thinking you've detected life on Mars and finding that it's *E. coli* from Pasadena," Kenneth Nealson, a University of Southern California geobiologist, said in 2001.

I can see Mars rising tonight, an ocher point in the dark. I ignore probability for a moment and imagine *E. coli* piggybacking on some early Martian probe—perhaps a Russian orbiter that lost control and crashed to the surface. *E. coli* would not take over the planet. In the cold, radioactive night, without a high-pressure atmosphere to push back against it, it would die. As I look at the ocher point, I think of Mars as a tiny failed colony of *E. coli* set against a vast, black petri dish. *Escherichia coli* helped guide us to an understanding of life on Earth. Now it scouts ahead, into the greater living universe.

ACKNOWLEDGMENTS

I am grateful to a number of scientists who have opened their labs, picked up their phones, and replied to my e-mails, all in order to teach me about *Escherichia coli.* They include Mark Achtman, Adam Arkin, M. Madan Babu, Steven Benner, Howard Berg, Mary Berlyn, Ronald Breaker, Sam Brown, George Church, Carol Cleland, James Collins, John Dennehy, Michael Doebeli, John Doyle, Michael Ellison, Thierry Emonet, Drew Endy, Thomas Ferenci, Finbarr Hayes, Peter Karp, Jay Keasling, Frank Keil, Andrew Knoll, Michael Krawinkel, Jan-Ulrich Kreft, Richard Lenski, Hirotada Mori, Kaare Nielsen, Christos Ouzounis, Mark Pallen, Bernhard Palsson, Arthur Pardee, Robert Pennock, Mark Ptashne, Margaret Riley, John Roth, Dean Rowe-Magnus, Jack Szostak, Phillip Tarr, Fred Tenover, Paul Thomas, Jeffrey Townsend, Paul Turner, David Ussery, Alexander van Oudenaarden, Barry Wanner, Daniel Weinreich, and George Williams.

I would also like to thank scientists and writers who looked over the manuscript or portions of it, including Mark Achtman, Uri Alon, Michael Balter, M. Madan Babu, Les Dethlefsen, Michael Feldgarden, Kevin Foster, James Hu, John Ingraham, Richard Lenski, Nicholas Matzke, Frederick Neidhardt, Monica Riley, and Eric Stewart. Moselio Schaechter was particularly generous with his time. Any errors that survived their careful scrutiny are entirely mine.

I wish to thank Doron Weber at the Alfred P. Sloan Foundation, which helped fund my research for this book. Thanks go also to my editors at the magazines and newspapers where I first wrote about some of the topics I revisit here: James Gorman and Erica Goode at *The New York Times,* Tim Appenzeller at *National Geographic,* David Grogan, Susan Kruglinski, and Corey Powell at *Discover,* Laura Helmuth at *Smithsonian,* Bruce Fellman and Kathrin Lassila at *Yale Alumni Magazine,* Leslie Roberts at *Science,* and Ricki Rusting at *Scientific American.*

My agent, Eric Simonoff, has never lost his fine power of discriminating between good book ideas and bad ones. When I saw him raise his eyebrows at my brief description of how *E. coli* swims, I realized I might have a good one. My thanks also go to Martin Asher, my editor at Pantheon, and Tadeusz Majewski, my illustrator.

Finally, there is my family. I thank my daughters, Charlotte and Veronica, for their indulgence while their father spent so much time writing about "the good germ." And my wife, Grace, provided the perfect blend of moral support and editorial criticism. Without her there would be no book. There would be no point.

NOTES

ONE: SIGNATURE

3 AND YET SCIENTISTS HAVE NO IDEA: Paul D. Thomas, personal communication.
3 THE OTHER 98 PERCENT: Bird, Stranger, and Dermitzakis, 2006.
4 THEY BELONG TO A SPECIES: Karp et al., 2007.

TWO: *E. COLI* AND THE ELEPHANT

6 IT THRIVED ON ALL MANNER OF FOOD: Dolman, 1970; Escherich, 1989.
7 "IT WOULD APPEAR TO BE A POINTLESS AND DOUBTFUL EXERCISE": Escherich, 1989, p. 352.
7 "WE MAY SAY IN PLAIN WORDS": Delbrück, 1969.
8 "FROM THE ELEPHANT TO BUTYRIC ACID BACTERIUM": quoted in Friedmann, 2004, p. 47.
8 BUT BACTERIA SUCH AS *E. COLI:* Brock, 1990; Judson, 1996.
10 MANY RESEARCHERS LOOKED AT BACTERIA: Brock, 1990.
11 "THE TERM 'GENE' CAN THEREFORE BE USED": quote from Gray and Tatum, 1944, p. 410; see also Tatum and Lederberg, 1947.
11 "THE LONG-SHOT GAMBLE": Lederberg, 1987, p. 26.
12 "*HOORAY*": Lederberg, 1946.
13 "BACTERIAL VIRUSES MAKE THEMSELVES KNOWN": quoted in Judson, 1996, p. 33.
13 THEY CALLED THEMSELVES THE PHAGE CHURCH: Stahl, 2001.
14 IT WAS SOMETHING CALLED DEOXYRIBONUCLEIC ACID: Avery, MacLeod, and McCarty, 1979.
15 "SO *STUPID* A SUBSTANCE": quoted in Judson, 1996, p. 40.
16 HERSHEY AND CHASE CONFIRMED HIS CONCLUSION: Hershey and Chase, 1952.
16 "A POWERFUL NEW PROOF": Watson, 1969, p. 119.
16 "THE UNTWIDDLING PROBLEM": quoted in Holmes, 2001, p. 78.
16 DELBRÜCK TRIED TO ANSWER THE QUESTION: Delbrück, 1954.
16 THE MOST BEAUTIFUL EXPERIMENT IN BIOLOGY: Meselson and Stahl, 1958.
18 AT THE CARNEGIE INSTITUTION: R. B. Roberts, 1955.
19 "AN ESSENTIALLY UNIVERSAL CODE": Marshall, Caskey, and Nirenberg, 1967, p. 826.

19 "TODAY, WE ARE LEARNING THE LANGUAGE": Clinton, 2000.

19 MANY BIOLOGISTS HAVE SPENT THEIR CAREERS: Echols, 2001; Neidhardt, 1996; Schaechter, Ingraham, and Neidhardt, 2006.

21 AN ENORMOUS PRESSURE INSIDE *E. COLI:* Norris et al., 2007.

22 TO UNCOVER ITS PATHWAYS: Sauer, Heinemann, and Zamboni, 2007.

23 *E. COLI* NEEDS IRON TO LIVE: Andrews, Robinson, and Rodriguez-Quinones, 2003; Wandersman and Delepelaire, 2004.

24 SUNLIGHT STRIKES THE PLANET: Michaelian, 2005.

25 THE FIRST SCIENTIST TO GET A GOOD LOOK: Berg, 2004.

26 WHERE THEY ACT LIKE A MICROBIAL TONGUE: Thiem, Kentner, and Sourjik, 2007.

27 ASTONISHINGLY TINY CHANGES IN THE CONCENTRATION OF MOLECULES: Bray, Levin, and Lipkow, 2007.

27 IT MAY BE MORE LIKE A BRAIN: M. D. Baker, Wolanin, and Stock, 2006.

27 WITH SOME LOOSE DNA TOSSED IN LIKE A BOWL OF TANGLED SPAGHETTI: Harold, 2005.

27 TO DIVIDE ALL OF LIFE INTO TWO GREAT GROUPS: Sapp, 2005.

28 HOW *E. COLI* ORGANIZES ITS DNA: Higgins, 2005; Thanbichler and Shapiro, 2006; Willenbrock and Ussery, 2004.

29 YET *E. COLI* CAN DO ALL OF THAT: O'Donnell, 2006.

29 TWO NEW CHROMOSOMES FORM: Jun and Mulder, 2006; Norris et al., 2007; Thanbichler and Shapiro, 2006; Woldringh and Nanninga, 2006.

29 A PROTEIN CALLED FTSZ: Bernhardt and de Boer, 2005; Goehring and Beckwith, 2005; Margolin, 2005.

31 INSTEAD, *E. COLI* SLAMS ON THE BRAKES: D. E. Chang, Smalley, and Conway, 2002; Higgins, 2005; Nystrom, 2004.

THREE: THE SYSTEM

32 ONE DAY IN JULY 1958: Jacob, 1995.

34 IT WOULD TAKE YEARS OF RESEARCH: Müller-Hill, 1996.

35 IN ANIMALS LIKE OURSELVES: Ben-Shahar et al., 2006.

36 SCIENTISTS HAVE CONTINUED TO PAY CLOSE ATTENTION: Alon, 2007.

36 IT ACTS AS A NOISE FILTER: Kalir, Mangan, and Alon, 2005.

39 FEED-FORWARD LOOPS ARE UNUSUALLY COMMON IN NATURE: Milo et al., 2002.

41 HE AND HIS COLLEAGUES BEGAN TO ANALYZE ITS HEAT-SHOCK PROTEINS: Kurata et al., 2006.

42 BERNHARD PALSSON, A BIOLOGIST AT THE UNIVERSITY OF CALIFORNIA, SAN DIEGO: Feist et al., 2007.

43 WHY DOES IT CHOOSE AMONG THE BEST FEW?: Trinh et al., 2006.

43 THE PICTURE THEY SEE: Ma and Zeng, 2003; Sauer, 2006.

43 THE BOW TIE ARCHITECTURE IN *E. COLI:* Csete and Doyle, 2004; Doyle and Csete, 2005; Doyle et al., 2005; Tanaka, Csete, and Doyle, 2005; Zhou, Carlson, and Doyle, 2005.

46 DANIEL KOSHLAND, A SCIENTIST AT THE UNIVERSITY OF CALIFORNIA, BERKELEY: Spudich and Koshland, 1976.

46 AARON NOVICK AND MILTON WEINER: Novick and Weiner, 1957.
47 SOME MICROBES WERE DARK: Elowitz et al., 2002.
47 THEY TURN OUT TO BE RESPONSIBLE: Ozbudak et al., 2004.
47 THEY SPEND MOST OF THEIR TIME SLIDING UP AND DOWN THE MICROBE'S DNA: Elf, Li, and Xie, 2007.
49 *E. COLI* WILL PULL METHYL GROUPS OFF ITS DNA: Lim and van Oudenaarden, 2007.
49 SOME OF THE FACTORS THAT SPIN THE WHEEL: Raser and O'Shea, 2005.
49 THE FIRST CLONED KITTEN, WHICH THEY NAMED CC: Shin et al., 2002.

FOUR: THE *E. COLI* WATCHER'S FIELD GUIDE

50 AN ISLAND VOLCANO CALLED KRAKATAU: For an excellent account of the history and ecology of this eruption, see Thornton, 1996.
51 TO MICROBES, A NEWBORN CHILD IS A KRAKATAU: Dethlefsen et al., 2006.
51 *E. COLI* IS A PIONEER: Wolfe, 2005.
52 "A ZEN-LIKE PHYSIOLOGY": J. W. Foster, 2004.
52 THE HAIRS BRING *E. COLI* TO A HALT: Thomas et al., 2004.
52 THE WARMTH OF THE GUT: White-Ziegler, Malhowski, and Young, 2007.
52 AT LEAST FOR A FEW DAYS: Favier et al., 2002; H. K. Park et al., 2005.
52 THIS ECOSYSTEM *E. COLI* HELPS TO BUILD: Dethlefsen et al., 2006.
53 *E. COLI* MAKES THE GUT RELIABLY COMFORTABLE: Jones et al., 2007.
53 WE, TOO, DEPEND ON OUR MICROBIAL JUNGLE: Backhed et al., 2005; Nicholson, Holmes, and Wilson, 2005.
54 IN 2003, JEFFRY STOCK AND HIS COLLEAGUES: S. Park et al., 2003.
55 SWARMING ALLOWS *E. COLI* TO GLIDE ACROSS A PETRI DISH: Inoue et al., 2007; Zorzano et al., 2005.
55 *E. COLI* CAN ALSO SETTLE DOWN: Beloin et al., 2004; Domka et al., 2007; Reisner et al., 2006.
55 A CLOUDY LAYER OF SCUM ON THEIR FLASKS: Ghannoum and O'Toole, 2004.
55 ON THE INNER WALLS OF OUR INTESTINES: Bollinger et al., 2007.
57 KNOWN AS COLICINS: Cascales et al., 2007.
57 THERE ARE MANY PREDATORS WAITING TO DEVOUR *E. COLI:* Meltz Steinberg and Levin, 2007.
57 OTHERS, SUCH AS THE BACTERIA *BDELLOVIBRIO:* Lambert et al., 2006.
57 THE BACTERIA *MYXOCOCCUS XANTHUS* RELEASE MOLECULES: Shi and Zusman, 1993.
58 LETS IT FEED ON MANY CARBON-BEARING MOLECULES—EVEN TNT: Stenuit et al., 2006.
58 IN AUSTRALIA, FOR EXAMPLE: Power et al., 2005.
59 *E. COLI* ALSO COMES IN FORMS THAT CAN SICKEN OR KILL: The best overall recent survey of pathogenic *E. coli* is Kaper, 2005.
59 JOHN BRAY, A BRITISH PATHOLOGIST: Bray, 1945.
60 WHEN SCIENTISTS COULD EXAMINE *SHIGELLA*'S GENES LETTER BY LETTER: Wirth et al., 2006.

60 SHIGELLA ALONE STRIKES 165 MILLION PEOPLE EVERY YEAR: Sansonetti, 2006.

60 FOR ALL ITS NOTORIETY: My account of E. coli O157:H7 is drawn from Elliott and Robins-Browne, 2005; Karch, Tarr, and Bielaszewska, 2005; Naylor, Gally, and Low, 2005; Pennington, 2003; Rangel et al., 2005; Tarr, Gordon, and Chandler, 2005; and Varma et al., 2003.

61 IN SEPTEMBER 2006, CONTAMINATED SPINACH: U.S. Department of Health and Human Services, Centers for Disease Control and Prevention, 2006.

62 ONCE THEY'VE FORMED A LARGE ENOUGH ARMY: Walters and Sperandio, 2006.

62 IT WANDERS: Jennison and Verma, 2004.

63 THEY OPEN UP MORE GAPS: Gorvel, 2006.

FIVE: EVERFLUX

65 "ONE MAY PERCEIVE": Lamarck, 1984, p. xxx.

67 AN ITALIAN REFUGEE SAT IN A COUNTRY CLUB: Luria, 1984; Luria and Delbrück, 1943.

67 THEY COLLABORATED WITH SCIENTISTS: Luria, Delbrück, and Anderson, 1943.

70 BUT WHEN LURIA AND DELBRÜCK FIRST PUBLISHED THE EXPERIMENT: Davis, 2003.

70 THE CONTROVERSY DID NOT DIE: Lederberg and Lederberg, 1952.

72 THEY SET OUT TO OBSERVE *E. COLI:* Zimmer, 2007b.

72 ONE OF THOSE SCIENTISTS WAS RICHARD LENSKI: The sources for my descriptions of Lenski's work include Crozat et al., 2005; Elena and Lenski, 2003; Lenski, 2003; Ostrowski, Rozen, and Lenski, 2005; Pelosi et al., 2006; Remold and Lenski, 2004; Rozen, Schneider, and Lenski, 2005; Schneider and Lenski, 2004; Travisano et al., 1995; Woods et al., 2006.

74 BERNHARD PALSSON AND HIS COLLEAGUES: Herring et al., 2006.

74 A ROUGHLY 1-IN-10,000 CHANCE: Perfeito et al., 2007.

75 "YOU PRESS THE REWIND BUTTON": Gould, 1989, p. 48.

77 IN THE EARLY 1990S, JULIAN ADAMS: Adams, 2004; Spencer et al., 2007.

77 MICHAEL DOEBELI AND HIS COLLEAGUES: Spencer et al., 2007.

78 IN LAKE APOYO: Barluenga et al., 2006.

SIX: DEATH AND KINDNESS

80 IN ORDER TO BE MORAL: Sapp, 1994, p. 21.

80 HIS ESSAYS WERE EVENTUALLY PUBLISHED: Kropotkin, 1919, p. 17.

82 YET ROBERTO KOLTER OF HARVARD AND A FORMER STUDENT: Vulić and Kolter, 2001.

83 GEORGE WILLIAMS, AN EVOLUTIONARY BIOLOGIST: Zimmer, 2004.

83 IN HIS 1966 BOOK, *ADAPTATION AND NATURAL SELECTION:* Williams, 1966.

83 ANOTHER YOUNG BIOLOGIST, WILLIAM HAMILTON: Segerstrale, in press.

84 *E. COLI* SUPPORTS THEIR VIEW OF LIFE: West et al., 2006.

84 CHEATERS CANNOT MARSHAL THESE DEFENSES: K. R. Foster, Parkinson, and Thompson, 2007.

84 JOAO XAVIER AND KEVIN FOSTER: Xavier and Foster, 2007.

85 CONFLICT AND COOPERATION STRIKE AN UNEASY BALANCE: Michod, 2007.

85 WE CALL THEIR SUCCESS CANCER: Zimmer, 2007a.

86 IT PAYS FOR THE POPULATION TO HEDGE ITS BETS: Mettetal et al., 2006; Wolf, Vazirani, and Arkin, 2005.

88 SCIENTISTS DISCOVERED SO-CALLED PERSISTER BACTERIA: Lewis, 2005, 2007.

88 A TEAM OF SCIENTISTS LED BY NATHALIE BALABAN: Balaban et al., 2004.

88 THAT'S THE THEORY OF KIM LEWIS: Lewis, 2007.

89 FOR THE ENTIRE POPULATION OF *E. COLI:* Kussell et al., 2005.

89 A NASTY SORT OF ALTRUISM: A. Gardner, West, and Buckling, 2004; West et al., 2006.

90 AS WITH PERSISTENCE: Mrak et al., 2007; Mulec et al., 2003.

90 SPITE, SOME EXPERIMENTS NOW SUGGEST: Kerr et al., 2002; Kirkup and Riley, 2004.

91 THE COMMON SIDE-BLOTCHED LIZARD: Sinervo, 2001.

93 SOME YEARS HE RAN A LITTLE FASTER: Williams, 1999.

93 SCIENTISTS STUDIED THE SOCKEYE SALMON: Morbey, Brassil, and Hendry, 2005.

94 ERIC STEWART, A MICROBIOLOGIST NOW AT NORTHEASTERN UNIVERSITY: Stewart et al., 2005.

96 IF IT SPENT ALL ITS RESOURCES ON REPAIR: Ackermann et al., 2007.

SEVEN: DARWIN AT THE DRUGSTORE

98 THE ERA OF ANTIBIOTICS BEGAN SUDDENLY: Levy, 2002; Salyers and Whitt, 2005.

99 IN 1948, THE YUGOSLAVIAN-BORN GENETICIST MILISLAV DEMEREC: Demerec, 1948.

99 TODAY THE WORLD CONSUMES: Wise and Soulsby, 2002.

99 MANY FARMERS TODAY PRACTICALLY DROWN THEIR ANIMALS: Graham, Boland, and Silbergeld, 2007.

101 TWO NEW MUTATIONS THAT MADE IT RESISTANT: Robicsek et al., 2006.

101 AFTER FIVE MONTHS AND TEN DIFFERENT ANTIBIOTICS: Rasheed et al., 1997; Tenover, 2006.

101 MICHAEL ZASLOFF, THEN A RESEARCH SCIENTIST: Shnayerson and Plotkin, 2002.

102 HE TEAMED UP WITH BELL: Perron, Zasloff, and Bell, 2006.

102 *E. COLI* AND OTHER BACTERIA ARE LOCKED IN AN EVOLUTIONARY RACE: Peschel and Sahl, 2006.

104 FLOYD ROMESBERG, A CHEMIST AT SCRIPPS RESEARCH INSTITUTE: Cirz et al., 2005.

105 JOHN CAIRNS, THEN AT HARVARD: Cairns, Overbaugh, and Miller, 1988.

106 SUSAN ROSENBERG OF BAYLOR COLLEGE OF MEDICINE: Ponder, Fonville, and Rosenberg, 2005.

106 NATURAL SELECTION, TENAILLON PROPOSES: Tenaillon, Denamur, and Matic, 2004.

107 ANTIBIOTICS MAY ALSO DRIVE THE RISE OF HIGH MUTATORS: Denamur et al., 2005.

107 WHEN NOTHING OF THE SORT HAS TAKEN PLACE: Roth et al., 2006.

108 SUSAN LINDQUIST OF THE WHITEHEAD INSTITUTE FOR BIOMEDICAL RESEARCH: Queitsch, Sangster, and Lindquist, 2002.

109 ICHIRO MATSUMURA, A BIOLOGIST AT EMORY UNIVERSITY: Patrick et al., 2007.

109 OTHER SPECIES MAY DEPEND ON THE SAME POTENTIAL: Francino, 2005.

109 SOME OF THESE EXTRA GENES: Myllykangas et al., 2006; Roth et al., 2006.

111 FEW SCIENTISTS OUTSIDE JAPAN: Watanabe, 1963.

111 IN A SURVEY OF *E. COLI* LIVING IN THE GREAT LAKES: Bartoloni et al., 2006.

EIGHT: OPEN SOURCE

113 AND IN HIS INTESTINES: Fricker, Spigelman, and Fricker, 1997.

114 LAWRENCE AND OCHMAN ESTIMATED: J. G. Lawrence and Ochman, 1998.

114 IN 2006, OCHMAN AND SEVERAL OTHER COLLEAGUES: Wirth et al., 2006.

115 SCIENTISTS COULD THEN COMPARE IT: Perna et al., 2001.

116 HORIZONTAL GENE TRANSFER MIGHT HAVE IMPORTED A FEW GENES: Wirth et al., 2006.

116 OF ALL THE *E. COLI* GENES SCIENTISTS HAD NOW IDENTIFIED: Welch et al., 2002.

117 THE LIST OF GENES SHARED: Binnewies et al., 2006; Chen et al., 2006.

117 IT'S UP TO 11,000 GENES NOW: David Ussery, personal communication.

117 VIRUSES IN THE OCEAN TRANSFER GENES TO NEW HOSTS: Sullivan and Baross, 2007.

118 OPEN-SOURCE EVOLUTION: Frost et al., 2005.

119 JUST AS IMPORTANT AS THE GENES *SHIGELLA* GAINED: Casalino et al., 2005.

119 SCIENTISTS WHO STUDY *E. COLI* O157:H7: Wick et al., 2005.

120 THEY SPECULATE THAT O157:H7'S TOXINS STIMULATE: Ferens, Cobbold, and Hovde, 2006.

120 WHEN PROTOZOANS ATTACK *E. COLI* COLONIES: Meltz Steinberg and Levin, 2007.

120 THEY FOUND GENES FOR CELL-KILLING FACTORS: Hejnova et al., 2005.

122 IT'S POSSIBLE THAT THE BACTERIA BENEFIT: Gamage et al., 2003, 2004, 2006; Shaikh and Tarr, 2003; Zhang et al., 2006.

123 ICHIZO KOBAYASHI, A GENETICIST AT THE UNIVERSITY OF TOKYO: Kobayashi, 2001.

NINE: PALIMPSEST

125 "THE COMPLETE GENOME SEQUENCE OF *E. COLI* K-12": Blattner et al., 1997.

126 MY FAVORITE IS AN OLD BATTERED BOOK: Chong, 2006.

128 CARL WOESE, A BIOLOGIST AT THE UNIVERSITY OF ILLINOIS: Woese and Fox, 1977.

130 W. FORD DOOLITTLE, A BIOLOGIST AT DALHOUSIE UNIVERSITY: Doolittle, 2000.

131 THE TREE OF LIFE STILL STANDS: Ge, Wang, and Kim, 2005.

131 HOWARD OCHMAN CAME TO THIS CONCLUSION: Lerat et al., 2005.

131 AT THE CENTER OF THE WHEEL IS THE LAST COMMON ANCESTOR: Ciccarelli et al., 2006.

132 CHRISTOS OUZOUNIS AND HIS COLLEAGUES: Ouzounis et al., 2006.

132 FOUR BILLION YEARS AGO, EARTH WAS REGULARLY DEVASTATED: Battistuzzi, Feijao, and Hedges, 2004.

133 OTHERS LIVE ON THE SIDES OF UNDERSEA VOLCANOES: Hou et al., 2004.

133 BUT SOME SPECIES, INCLUDING THE ANCESTORS OF *E. COLI:* Raymond and Segre, 2006.

133 TODAY *E. COLI* CAN STILL SWITCH BACK AND FORTH: Tomitani et al., 2006.

134 TODAY BACTERIA HAVE AN IMPRESSIVE RANGE OF DEFENSES: Matz and Kjelleberg, 2005; Matz et al., 2005.

134 THE FEDERAL COURTHOUSE IN HARRISBURG: My account of *Kitzmiller v. Dover* is based on several sources, including Humes, 2007; Talbot, 2005; and TalkOrigins Archive, 2006.

136 "WE WOULD PREDICT THAT WE'D SEE": Bliss, 1981.

137 "TO THE MICROMECHANICIANS OF INDUSTRIAL RESEARCH AND DEVELOPMENT OPERATIONS": Lumsden, 1994.

137 BIOLOGISTS AND MATHEMATICIANS ALIKE: Van Till, 2002.

138 "SUCH SYSTEMS SIMPLY DEFY DARWINIST EXPLANATIONS": Hartwig, 2002.

140 "A SINGLE SYSTEM COMPOSED OF SEVERAL WELL-MATCHED, INTERACTING PARTS": quotations from Behe, 1996, p. 39.

143 "IF IT LOOKS DESIGNED, MAYBE IT IS": quoted in "Intelligently Designed Apparel and Merchandise," CafePress.com, http://www.cafepress.com/accessresearch/982234; accessed June 27, 2007.

144 IN 2005, MARK PALLEN OF THE UNIVERSITY OF BIRMINGHAM: Ren et al., 2005.

145 PALLEN AND NICHOLAS MATZKE: Pallen and Matzke, 2006.

146 *E. COLI'S* CONTROL NETWORKS HAVE AN ANCIENT HISTORY OF THEIR OWN: Cosentino Lagomarsino et al., 2007.

146 IN 2006, M. MADAN BABU, A BIOLOGIST AT THE UNIVERSITY OF CAMBRIDGE: Babu and Aravind, 2006.

147 ONE OF THE SIMPLEST MEANS: Zinser and Kolter, 2004.

148 *E. COLI'S* NETWORK GREW IN A SIMILAR WAY: Zhou, Carlson, and Doyle, 2005.

151 "THE RNA WORLD": Gesteland, Cech, and Atkins, 2006.

151 IN THE 1990S, RONALD BREAKER, A BIOCHEMIST AT YALE: Barrick and Breaker, 2007.

154 HE PROPOSES THAT THE PROFOUND SPLIT: Zimmer, 2006.

TEN: PLAYING NATURE

157 A $75 BILLION INDUSTRY: S. Lawrence, 2007b.

157 NOW 250 MILLION ACRES OF FARMLAND: S. Lawrence, 2007a.

158 A NEW KIND OF GENETIC ENGINEERING CALLED SYNTHETIC BIOLOGY: Baker et al., 2006.

158 BIOTECHNOLOGY WAS BORN MANY TIMES: Bud, 1993.

159 THE FIRST NEOLITHIC BIOTECHNOLOGISTS WERE MANIPULATING MICROBES: Mira, Pushker, and Rodriguez-Valera, 2006.

160 BY BOMBARDING THE MOLD THAT MAKES PENICILLIN: Adrio and Demain, 2006.

160 J.B.S. HALDANE INDULGED IN SOME SCIENCE FICTION: Haldane, 1923.

161 AND THOSE TOOLS WOULD EVENTUALLY BE USED: My account of the history of genetic engineering is based mainly on Hall, 2002; Jackson and Stich, 1979; Krimsky, 1982; Rogers, 1977; Singer, 2001; Wade, 1977; Watson and Tooze, 1981; Wright, 1994; and Zilinskas and Zimmerman, 1986.

161 SOMEONE HAD ISOLATED GENES: Shapiro et al., 1969.

161 "THE STEPS DO NOT EXIST NOW": quoted in Reinhold, 1969.

163 "ARE WE WISE ENOUGH TO BE TAMPERING": quoted in Krimsky, 1982, p. 35.

163 "I DIDN'T WANT TO BE THE PERSON": quoted in Krimsky, 1982, p. 31.

165 "NEW STRAINS OF LIFE—OR DEATH": Cavalieri, 1976.

165 "IT WAS NEVER THE INTENTION": Watson and Tooze, 1981, p. ix.

166 "SCIENTISTS ALONE DECIDED TO IMPOSE A MORATORIUM": quoted in Jackson and Stich, 1979, p. 101.

166 "BY THE YEAR 2000 VIRTUALLY ALL THE MAJOR HUMAN DISEASES": quoted in Dutton, Preston, and Pfund, 1988, p. 200.

166 IN 1977, THE NATIONAL ACADEMY OF SCIENCES HELD A PUBLIC FORUM: National Academy of Sciences, 1977.

167 "FROM THE POINT OF PUBLIC HEALTH": quoted from Cavalieri, 1976.

167 "AN IRREVERSIBLE ATTACK ON THE BIOSPHERE": Chargaff and Simring, 1976.

167 "NIGHTMARISH AND DISASTROUS": Berg and Cohen quoted in Watson and Tooze, 1981, p. 389.

167 "WE WERE JACKASSES": Watson and Tooze, 1981, p. 437.

167 "I'M AFRAID THAT BY CRYING WOLF": Watson and Tooze, 1981, p. 155.

169 HUMULIN, ITS MICROBE-PRODUCED INSULIN: Figure from Eli Lilly and Company, "Humulin," http://www.lillydiabetes.com/product/humulin_family.jsp?reqNavId=5.3; accessed June 27, 2007.

171 "THE FIRST SYNTHETIC LIFE FORM": Service, 2003, p. 640.

171 SCIENTISTS HAVE ADDED OVER THIRTY MORE UNNATURAL ACIDS: Wang, Xie, and Schultz, 2006.

173 MICHAEL ELOWITZ AT CALIFORNIA INSTITUTE OF TECHNOLOGY: Elowitz and Leibler, 2000.

173 THE SECOND REPORT CAME FROM THE LABORATORY OF JAMES COLLINS: T. S. Gardner, Cantor, and Collins, 2000.

174 IN 2004, STUDENTS AT THE UNIVERSITY OF TEXAS AND THE UNIVERSITY OF CALIFORNIA, SAN FRANCISCO: Levskaya et al., 2005.

175 THE DRUG, KNOWN AS ARTEMISININ: M. C. Chang and Keasling, 2006.

176 SINCE 2001, DREW ENDY AND THOMAS KNIGHT OF MIT HAVE BEEN BUILDING: Endy, 2005.

176 GEORGE CHURCH AND HIS COLLEAGUES HAVE DRAWN UP A LIST: Forster and Church, 2006.

176 ALBERT LIBCHABER TOOK AN EVEN SIMPLER APPROACH: Noireaux et al., 2005.

177 "BIOTECH HAS ALREADY IGNITED WORLDWIDE PROTESTS": quoted in ETC Group, 2006.

178 THEY FOUND THAT THE BACTERIA RAPIDLY DISAPPEARED: Bogosian and Kane, 1991; Heitkamp et al., 1993.

178 STUDIES SUGGEST THAT EVEN IF AN ALIEN GENE GAVE BACTERIA A COMPETITIVE ADVANTAGE: Nielsen and Townsend, 2004.

180 THE INCIDENCE OF TYPE 2 DIABETES HAS DOUBLED: Statistics from U.S. Department of Health and Human Services, Centers for Disease Control and Prevention, "Diabetes Data & Trends," http://www.cdc.gov/diabetes/statistics/prev/national/figpersons.htm; accessed June 27, 2007.

180 CASES OF DIABETES WORLDWIDE HAVE INCREASED TENFOLD: World Health Organization, Media Centre, "Diabetes," http://www.who/int/mediacentre/factsheets/fs312/en/'; accessed June 27, 2007.

180 *E. COLI* HAS PROVIDED INSULIN: Eli Lilly Company, "Humulin," http://www.lillydiabetes.com/product/humulin_family.jsp?reqNavId=5.3; accessed June 27, 2007.

180 THAT SUGAR COMES INCREASINGLY FROM HIGH-FRUCTOSE CORN SYRUP: Crabb and Shetty, 1999.

180 THE COMPANY HAD TO PAY $50 MILLION TO SETTLE CHARGES: Nordenberg, 1999.

180 WE MUST RESIST EMPTY FEAR AND EMPTY HYPE: Avise, 2004.

181 "THE AFFLUENT NATIONS CAN AFFORD": Borlaug, 2000, p. 489.

181 GOLDEN RICE, A STRAIN OF RICE ENGINEERED: Babili and Beyer, 2005.

181 "THIS RICE COULD SAVE A MILLION KIDS A YEAR": Nash, 2000.

181 "IN FIGHTING AGAINST 'GOLDEN RICE' REACHING THE POOR": Potrykus, 2001, p. 1160.

182 IT MAY NOT BRING MUCH BENEFIT AT ALL: Krawinkel, 2007.

182 VITAMIN A HAS TO BE CONSUMED ALONG WITH DIETARY FAT: Schnapp and Schiermeier, 2001.

182 ABOUT 80 PERCENT OF ALL THE TRANSGENIC CROPS: Service, 2007.

183 STUDIES INDICATED THAT FARMERS WHO GREW THE TRANSGENIC CROPS: Raney, 2006.

183 FARMERS BEGAN TO NOTICE HORSEWEED AND MORNING GLORY: Owen and Zelaya, 2005.

183 AND ONCE THE WEEDS EVOLVED THEIR RESISTANCE: Zelaya, Owen, and VanGessel, 2007.

183 THEY HAD ENGINEERED PLANTS WITH GENES THAT MAKE THEM RESISTANT: Behrens et al., 2007.

183 FARMERS CAN RESORT TO OLD-FASHIONED METHODS: Sandermann, 2006.

184 "WE CAN NOW TRANSFORM THAT EVOLUTIONARY TREE": quoted in Jackson and Stich, 1979, p. 97.

185 THE MIT BIOLOGIST JONATHAN KING DECLARED: quoted in Hall, 2002, p. 51.

185 "THE GENETIC STRUCTURES OF LIVING BEINGS": Silver, 2006, p. 287.

185 "LET THE EARTH BRING FORTH GRASS": Gen. 1:11–24 (King James version).

186 "THOU SHALT NOT LET THY CATTLE": Lev. 19:19 (King James version).

186 "NEITHER SHALT THOU LIE": Lev. 18:23 (King James version).

186 "USING HUMAN PROCREATION": Jones, 2006.

186 "REPUGNANCE MAY BE THE ONLY VOICE LEFT": Kass and Wilson, 1998, p. 19.

186 PRESIDENT BUSH CALLED FOR A BAN: Bush, 2006.

186 "RESPECT FOR HUMAN DIGNITY AND THE INTEGRITY OF THE HUMAN SPECIES": U.S. Congress, 2005.

187 "A THING EITHER IS OR IS NOT A WHOLE HUMAN BEING": George and Gomez-Lobo, 2005, p. 202.

187 *E. COLI* STARVES AND SUFFERS: Chou, 2007.

187 EACH KIND HAS ITS OWN ESSENCE: Gelman, 2004.

188 "THE THING IS AN ABOMINATION": Wells, 1896.

189 CONSIDER A GENE CALLED MICROCEPHALIN: Evans et al., 2006.

191 "IN A WORLD WHOSE ONCE-GIVEN NATURAL BOUNDARIES": Kass, 1997.

191 AND SOME CHIMERAS WILL PROBABLY BE BANNED: Scott, 2006.

ELEVEN: *N* EQUALS 1

194 "BUT ITS DOMAIN HAS BEEN LIMITED": Lederberg, 1963.

195 "INVASION FROM MARS? MICROBES!": *Los Angeles Examiner,* March 20, 1960.

195 "THESE QUESTIONS MIGHT BE ANSWERED IN TWO WAYS": Lederberg, 1963.

195 "WE CAN DEFER OUR CONCERN": Lederberg, 1960, p. 394.

195 NASA AGREED: Dick, 1998.

196 "THAT'S THE BALL GAME": quoted in Dick and Strick, 2004.

196 "WE CAN NO LONGER BE CONFIDENT": quoted in McElheny, 1976.

197 SYNTHETIC BIOLOGY BLURS INTO ASTROBIOLOGY: Benner, Ricardo, and Carrigan, 2004.

198 PLANETS MAY TRADE BITS OF THEMSELVES: Gladman et al., 2005; Warmflash and Weiss, 2005.

199 THE MOST EXTREME FORM OF PANSPERMIA: Napier, 2004.

199 IT FLOATS INSIDE THE BODIES OF THE ASTRONAUTS: La Duc et al., 2004.

199 "THE FIELD IS HAUNTED BY THINKING": quoted in Clarke, 2001, p. 248.

SELECTED BIBLIOGRAPHY

Ackermann, M., L. Chao, C. T. Bergstrom, and M. Doebeli. 2007. On the evolutionary origin of aging. *Aging Cell* 6 (2):235–44.

Adams, J. 2004. Microbial evolution in laboratory environments. *Res Microbiol* 155 (5):311–18.

Adrio, J. L., and A. L. Demain. 2006. Genetic improvement of processes yielding microbial products. *FEMS Microbiol Rev* 30 (2):187–214.

Alon, U. 2007. Network motifs: Theory and experimental approaches. *Nat Rev Genet* 8 (6):450–61.

Andrews, S. C., A. K. Robinson, and F. Rodriguez-Quinones. 2003. Bacterial iron homeostasis. *FEMS Microbiol Rev* 27 (2–3):215–37.

Avery, O. T., C. M. MacLeod, and M. McCarty. 1979. Studies on the chemical nature of the substance inducing transformation of pneumococcal types: Induction of transformation by a desoxyribonucleic acid fraction isolated from pneumococcus type III. *J Exp Med* 149 (2):297–326.

Avise, John C. 2004. *The hope, hype, and reality of genetic engineering: Remarkable stories from agriculture, industry, medicine, and the environment.* New York: Oxford University Press.

Babili, S. al-, and P. Beyer. 2005. Golden rice—five years on the road—five years to go? *Trends Plant Sci* 10 (12):565–73.

Babu, M. M., and L. Aravind. 2006. Adaptive evolution by optimizing expression levels in different environments. *Trends Microbiol* 14 (1):11–14.

Backhed, F., R. E. Ley, J. L. Sonnenburg, D. A. Peterson, and J. I. Gordon. 2005. Host-bacterial mutualism in the human intestine. *Science* 307 (5717):1915–20.

Baker, D., G. Church, J. Collins, D. Endy, J. Jacobson, J. Keasling, P. Modrich, C. Smolke, and R. Weiss. 2006. Engineering life: Building a fab for biology. *Sci Am* 294 (6):44–51.

Baker, M. D., P. M. Wolanin, and J. B. Stock. 2006. Signal transduction in bacterial chemotaxis. *Bioessays* 28 (1):9–22.

Balaban, N. Q., J. Merrin, R. Chait, L. Kowalik, and S. Leibler. 2004. Bacterial persistence as a phenotypic switch. *Science* 305 (5690):1622–25.

Barluenga, M., K. N. Stolting, W. Salzburger, M. Muschick, and A. Meyer. 2006. Sympatric speciation in Nicaraguan crater lake cichlid fish. *Nature* 439 (7077):719–23.

Barrick, J. E., and R. R. Breaker. 2007. The power of riboswitches. *Sci Am* 296 (1):50–57.

Bartoloni, A., L. Pallecchi, M. Benedetti, C. Fernandez, Y. Vallejos, E. Guzman, A. L. Villagran, A. Mantella, C. Lucchetti, F. Bartalesi, M. Strohmeyer, A. Bechini, H. Gamboa, H. Rodriguez, T. Falkenberg, G. Kronvall, E. Gotuzzo, F. Paradisi, and G. M. Rossolini. 2006. Multidrug-resistant commensal *Escherichia coli* in children, Peru and Bolivia. *Emerg Infect Dis* 12 (6):907–13.

Battistuzzi, F. U., A. Feijao, and S. B. Hedges. 2004. A genomic timescale of prokaryote evolution: Insights into the origin of methanogenesis, phototrophy, and the colonization of land. *BMC Evol Biol* 4:44.

Behe, Michael. 1996. *Darwin's black box.* New York: Free Press.

Behrens, Mark R., Nedim Mutlu, Sarbani Chakraborty, Razvan Dumitru, Wen Zhi Jiang, Bradley J. LaVallee, Patricia L. Herman, Thomas E. Clemente, and Donald P. Weeks. 2007. Dicamba resistance: Enlarging and preserving biotechnology-based weed management strategies. *Science* 316 (5828):1185–88.

Beloin, C., J. Valle, P. Latour-Lambert, P. Faure, M. Kzreminski, D. Balestrino, J. A. Haagensen, S. Molin, G. Prensier, B. Arbeille, and J. M. Ghigo. 2004. Global impact of mature biofilm lifestyle on *Escherichia coli* K-12 gene expression. *Mol Microbiol* 51 (3):659–74.

Benner, S. A., A. Ricardo, and M. A. Carrigan. 2004. Is there a common chemical model for life in the universe? *Curr Opin Chem Biol* 8 (6):672–89.

Ben-Shahar, Y., K. Nannapaneni, T. L. Casavant, T. E. Scheetz, and M. J. Welsh. 2006. Eukaryotic operon-like transcription of functionally related genes in *Drosophila. Proc Natl Acad Sci USA* 104 (1):222–27.

Berg, Howard C. 2004. E. coli *in motion.* Biological and Medical Physics Series. New York: Springer.

Bernhardt, T. G., and P. A. de Boer. 2005. SlmA, a nucleoid-associated, FtsZ binding protein required for blocking septal ring assembly over chromosomes in *E. coli. Mol Cell* 18 (5):555–64.

Binnewies, T. T., Y. Motro, P. F. Hallin, O. Lund, D. Dunn, T. La, D. J. Hampson, M. Bellgard, T. M. Wassenaar, and D. W. Ussery. 2006. Ten years of bacterial genome sequencing: Comparative-genomics-based discoveries. *Funct Integr Genomics* 6 (3):165–85.

Bird, C. P., B. E. Stranger, and E. T. Dermitzakis. 2006. Functional variation and evolution of non-coding DNA. *Curr Opin Genet Dev* 16 (6):559–64.

Blattner, F. R., G. Plunkett III, C. A. Bloch, N. T. Perna, V. Burland, M. Riley, J. Collado-Vides, J. D. Glasner, C. K. Rode, G. F. Mayhew, J. Gregor, N. W. Davis, H. A. Kirkpatrick, M. A. Goeden, D. J. Rose, B. Mau, and Y. Shao. 1997. The complete genome sequence of *Escherichia coli* K-12. *Science* 277 (5331):1453–74.

Bliss, Richard. 1981. Creation science papers. Series 1, subseries 1, box 4, folder 5. Special collections, University of Arkansas Libraries, Fayetteville.

Bogosian, G., and J. F. Kane. 1991. Fate of recombinant *Escherichia coli* K-12 strains in the environment. *Adv Appl Microbiol* 36:87–131.

Bollinger, R. R., A. S. Barbas, E. L. Bush, S. S. Lin, and W. Parker. 2007. Biofilms in the normal human large bowel: Fact rather than fiction. *Gut* 56:1481–12.

Borlaug, N. E. 2000. Ending world hunger: The promise of biotechnology and the threat of antiscience zealotry. *Plant Physiol* 124 (2):487–90.

Bray, D., M. D. Levin, and K. Lipkow. 2007. The chemotactic behavior of computer-based surrogate bacteria. *Curr Biol* 17 (1):12–19.

Bray, J. 1945. Isolation of antigenically homogeneous strains of *Bact. coli neapolitanum* from summer diarrhea of infants. *J Pathol Bacteriol* 57:239–47.

Brock, Thomas D. 1990. *The emergence of bacterial genetics.* Cold Spring Harbor, N.Y.: Cold Spring Harbor Laboratory Press.

Bud, Robert. 1993. *The uses of life: A history of biotechnology.* New York: Cambridge University Press.

Bush, George. 2006. State of the Union Address. White House. http://www.whitehouse.gov/stateoftheunion/2006/.

Cairns, J., J. Overbaugh, and S. Miller. 1988. The origin of mutants. *Nature* 335 (6186):142–45.

Casalino, M., M. C. Latella, G. Prosseda, P. Ceccarini, F. Grimont, and B. Colonna. 2005. Molecular evolution of the lysine decarboxylase-defective phenotype in *Shigella sonnei. Int J Med Microbiol* 294 (8):503–12.

Cascales, E., S. K. Buchanan, D. Duche, C. Kleanthous, R. Lloubes, K. Postle, M. Riley, S. Slatin, and D. Cavard. 2007. Colicin biology. *Microbiol Mol Biol Rev* 71 (1):158–229.

Cavalieri, L. F. 1976. New strains of life—or death. *New York Times Magazine,* August 22, 1976.

Chang, D. E., D. J. Smalley, and T. Conway. 2002. Gene expression profiling of *Escherichia coli* growth transitions: An expanded stringent response model. *Mol Microbiol* 45 (2):289–306.

Chang, M. C., and J. D. Keasling. 2006. Production of isoprenoid pharmaceuticals by engineered microbes. *Nat Chem Biol* 2 (12):674–81.

Chargaff, E., and F. R. Simring. 1976. On the dangers of genetic meddling. *Science* 192 (4243):938.

Chen, S. L., C. S. Hung, J. Xu, C. S. Reigstad, V. Magrini, A. Sabo, D. Blasiar, T. Bieri, R. R. Meyer, P. Ozersky, J. R. Armstrong, R. S. Fulton, J. P. Latreille, J. Spieth, T. M. Hooton, E. R. Mardis, S. J. Hultgren, and J. I. Gordon. 2006. Identification of genes subject to positive selection in uropathogenic strains of *Escherichia coli:* A comparative genomics approach. *Proc Natl Acad Sci USA* 103 (15):5977–82.

Chong, Jia-Rui. 2006. 13th century text hides words of Archimedes. *Los Angeles Times,* December 26, 2006.

Chou, C. P. 2007. Engineering cell physiology to enhance recombinant protein production in *Escherichia coli. Appl Microbiol Biotechnol,* in press.

Ciccarelli, Francesca D., Tobias Doerks, Christian von Mering, Christopher J. Creevey, Berend Snel, and Peer Bork. 2006. Toward automatic reconstruction of a highly resolved tree of life. *Science* 311 (5765):1283–87.

Cirz, R. T., J. K. Chin, D. R. Andes, V. de Crécy-Lagard, W. A. Craig, and F. E. Romesberg. 2005. Inhibition of mutation and combating the evolution of antibiotic resistance. *PLoS Biol* 3 (6):e176.

Clarke, T. 2001. The stowaways. *Nature* 413 (6853):247–48.

Clinton, W. J. 2000. Speech by President on Completion of First Survey of Entire Human Genome. http://www.clintonpresidentialcenter.org; accessed October 24, 2007.

Cosentino Lagomarsino, M., P. Jona, B. Bassetti, and H. Isambert. 2007. Hierarchy and feedback in the evolution of the *Escherichia coli* transcription network. *Proc Natl Acad Sci USA* 104 (13):5516–20.

Crabb, W. D., and J. K. Shetty. 1999. Commodity scale production of sugars from starches. *Curr Opin Microbiol* 2 (3):252–56.

Crozat, E., N. Philippe, R. E. Lenski, J. Geiselmann, and D. Schneider. 2005. Long-term experimental evolution in *Escherichia coli.* XII. DNA topology as a key target of selection. *Genetics* 169 (2):523–32.

Csete, M., and J. Doyle. 2004. Bow ties, metabolism and disease. *Trends Biotechnol* 22 (9):446–50.

Davis, Rowland H. 2003. *The microbial models of molecular biology: From genes to genomes.* New York: Oxford University Press.

Delbrück, M. 1954. On the replication of desoxyribonucleic acid (DNA). *Proc Natl Acad Sci USA* 40:783–88.

———. 1969. Nobel lecture. http://nobelprize.org/nobel_prizes/medicine/laureates/1969/delbruck-lecture.html.

Demerec, M. 1948. Origin of bacterial resistance to antibiotics. *J Bacteriol* 56 (1):63–74.

Denamur, E., O. Tenaillon, C. Deschamps, D. Skurnik, E. Ronco, J. L. Gaillard, B. Picard, C. Branger, and I. Matic. 2005. Intermediate mutation frequencies favor evolution of multidrug resistance in *Escherichia coli. Genetics* 171 (2):825–27.

Dethlefsen, L., P. B. Eckburg, E. M. Bik, and D. A. Relman. 2006. Assembly of the human intestinal microbiota. *Trends Ecol Evol* 21 (9):517–23.

Dick, Steven J. 1998. *Life on other worlds: The twentieth-century extraterrestrial life debate.* New York: Cambridge University Press.

Dick, Steven J., and James Edgar Strick. 2004. *The living universe: NASA and the development of astrobiology.* New Brunswick, N.J.: Rutgers University Press.

Dolman, Claude E. 1970. Theodor Escherich. In *Dictionary of scientific biography,* ed. C. C. Gillispie. New York: Scribner.

Domka, J., J. Lee, T. Bansal, and T. K. Wood. 2007. Temporal gene-expression in *Escherichia coli* K-12 biofilms. *Environ Microbiol* 9 (2):332–46.

Doolittle, W. F. 2000. Uprooting the tree of life. *Sci Am* 282 (2):90–95.

Doyle, J., D. L. Alderson, L. Li, S. Low, M. Roughan, S. Shalunov, R. Tanaka, and W. Willinger. 2005. The "robust yet fragile" nature of the Internet. *Proc Natl Acad Sci USA* 102 (41):14497–502.

Doyle, J., and M. Csete. 2005. Motifs, control, and stability. *PLoS Biol* 3 (11):e392.

Dutton, Diana Barbara, Thomas A. Preston, and Nancy E. Pfund. 1988. *Worse than the disease: Pitfalls of medical progress.* New York: Cambridge University Press.

Echols, Harrison. 2001. *Operators and promoters: The story of molecular biology and its creators.* Ed. Carol Gross. Berkeley: University of California Press.

Elena, S. F., and R. E. Lenski. 2003. Evolution experiments with microorganisms: The dynamics and genetic bases of adaptation. *Nat Rev Genet* 4 (6):457–69.

Elf, J., G. W. Li, and X. S. Xie. 2007. Probing transcription factor dynamics at the single-molecule level in a living cell. *Science* 316 (5828):1191–94.

Elliott, E. J., and R. M. Robins-Browne. 2005. Hemolytic uremic syndrome. *Curr Probl Pediatr Adolesc Health Care* 35 (8):310–30.

Elowitz, M. B., and S. Leibler. 2000. A synthetic oscillatory network of transcriptional regulators. *Nature* 403 (6767):335–38.

Elowitz, M. B., A. J. Levine, E. D. Siggia, and P. S. Swain. 2002. Stochastic gene expression in a single cell. *Science* 297 (5584):1183–86.

Endy, D. 2005. Foundations for engineering biology. *Nature* 438 (7067):449–53.

Escherich, T. 1989. The intestinal bacteria of the neonate and breast-fed infant. 1885. *Rev Infect Dis* 11 (2):352–56.

ETC Group. 2006. Global coalition sounds the alarm on synthetic biology, demands oversight and societal debate. Press release. http://www.etcgroup.org/upload/publication/8/01/nr_synthetic_bio_19th_may_2006.pdf.

Evans, P. D., N. Mekel-Bobrov, E. J. Vallender, R. R. Hudson, and B. T. Lahn. 2006. Evidence that the adaptive allele of the brain size gene microcephalin introgressed into *Homo sapiens* from an archaic *Homo* lineage. *Proc Natl Acad Sci USA* 103 (48):18178–83.

Favier, C. F., E. E. Vaughan, W. M. De Vos, and A. D. Akkermans. 2002. Molecular monitoring of succession of bacterial communities in human neonates. *Appl Environ Microbiol* 68 (1):219–26.

Feist, Adam M., Christopher S. Henry, Jennifer L. Reed, Markus Krummenacker, Andrew R. Joyce, Peter D. Karp, Linda J. Broadbelt, Vassily Hatzimanikatis, and Bernhard O. Palsson. 2007. A genome-scale metabolic reconstruction for *Escherichia coli* K-12 MG1655 that accounts for 1260 ORFs and thermodynamic information. *Mol Syst Biol* 3:121.

Ferens, W. A., R. Cobbold, and C. J. Hovde. 2006. Intestinal Shiga toxin–producing *Escherichia coli* bacteria mitigate bovine leukemia virus infection in experimentally infected sheep. *Infect Immun* 74 (5):2906–16.

Forster, A. C., and G. M. Church. 2006. Towards synthesis of a minimal cell. *Mol Syst Biol* 2:45.

Foster, J. W. 2004. *Escherichia coli* acid resistance: Tales of an amateur acidophile. *Nat Rev Microbiol* 2 (11):898–907.

Foster, K. R., K. Parkinson, and C. R. Thompson. 2007. What can microbial genetics teach sociobiology? *Trends Genet* 23 (2):74–80.

Francino, M. P. 2005. An adaptive radiation model for the origin of new gene functions. *Nat Genet* 37 (6):573–77.

Fricker, E. J., M. Spigelman, and C. R. Fricker. 1997. The detection of *Escherichia coli* DNA in the ancient remains of Lindow Man using the polymerase chain reaction. *Lett Appl Microbiol* 24 (5):351–54.

Friedmann, H. C. 2004. From *Butyribacterium* to *E. coli:* An essay on unity in biochemistry. *Perspect Biol Med* 47 (1):47–66.

Frost, L. S., R. Leplae, A. O. Summers, and A. Toussaint. 2005. Mobile genetic elements: The agents of open source evolution. *Nat Rev Microbiol* 3 (9):722–32.

Gamage, S. D., A. K. Patton, J. F. Hanson, and A. A. Weiss. 2004. Diversity and host range of Shiga toxin–encoding phage. *Infect Immun* 72 (12):7131–39.

Gamage, S. D., A. K. Patton, J. E. Strasser, C. L. Chalk, and A. A. Weiss. 2006. Commensal bacteria influence *Escherichia coli* O157:H7 persistence and Shiga toxin production in the mouse intestine. *Infect Immun* 74 (3):1977–83.

Gamage, S. D., J. E. Strasser, C. L. Chalk, and A. A. Weiss. 2003. Nonpathogenic *Escherichia coli* can contribute to the production of Shiga toxin. *Infect Immun* 71 (6):3107–15.

Gardner, A., S. A. West, and A. Buckling. 2004. Bacteriocins, spite and virulence. *Proc Biol Sci* 271 (1547):1529–35.

Gardner, T. S., C. R. Cantor, and J. J. Collins. 2000. Construction of a genetic toggle switch in *Escherichia coli. Nature* 403 (6767):339–42.

Ge, F., L. S. Wang, and J. Kim. 2005. The cobweb of life revealed by genome-scale estimates of horizontal gene transfer. *PLoS Biol* 3 (10):e316.

Gelman, S. A. 2004. Psychological essentialism in children. *Trends Cogn Sci* 8 (9):404–9.

George, R. P., and A. Gomez-Lobo. 2005. The moral status of the human embryo. *Perspect Biol Med* 48 (2):201–10.

Gesteland, Raymond F., Thomas Cech, and John F. Atkins. 2006. *The RNA world: The nature of modern RNA suggests a prebiotic RNA.* 3rd ed. Cold Spring Harbor Monograph Series 43. Cold Spring Harbor, N.Y.: Cold Spring Harbor Laboratory Press.

Ghannoum, Mahmoud A., and George A. O'Toole. 2004. *Microbial biofilms.* Washington, D.C.: ASM Press.

Gladman, B., L. Dones, H. F. Levison, and J. A. Burns. 2005. Impact seeding and reseeding in the inner solar system. *Astrobiology* 5 (4):483–96.

Goehring, N. W., and J. Beckwith. 2005. Diverse paths to midcell: Assembly of the bacterial cell division machinery. *Curr Biol* 15 (13):R514–26.

Gorvel, J. P. 2006. Microbiology: Bacterial bushwacking through a microtubule jungle. *Science* 314 (5801):931–32.

Gould, Stephen Jay. 1989. *Wonderful life: The Burgess Shale and the nature of history.* New York: W. W. Norton.

Graham, J. P., J. J. Boland, and E. Silbergeld. 2007. Growth-promoting antibiotics in food animal production: An economic analysis. *Public Health Reports* 122 (1):79–87.

Gray, C. H., and E. L. Tatum. 1944. X-ray induced growth factor requirements in bacteria. *Proc Natl Acad Sci USA* 30 (12):404–10.

Haldane, J.B.S. 1923. *Daedalus; or, Science and the future: A paper read to the Heretics, Cambridge, on February 4th, 1923.* http://www.cscs.umich.edu/~crshalizi/Daedalus .html; accessed October 24, 2007.

Hall, Stephen S. 2002. *Invisible frontiers: The race to synthesize a human gene.* New York: Oxford University Press.

Harold, F. M. 2005. Molecules into cells: Specifying spatial architecture. *Microbiol Mol Biol Rev* 69 (4):544–64.

Hartwig, Mark. 2002. Whose comfortable myth? *Focus on the Family,* June 2002.

Heitkamp, M. A., J. F. Kane, P. J. Morris, M. Bianchini, M. D. Hale, and G. Bogosian. 1993. Fate in sewage of a recombinant *Escherichia coli* K-12 strain used in the commercial production of bovine somatotropin. *J Ind Microbiol* 11 (4):243–52.

Hejnova, J., U. Dobrindt, R. Nemcova, C. Rusniok, A. Bomba, L. Frangeul, J. Hacker, P. Glaser, P. Sebo, and C. Buchrieser. 2005. Characterization of the flexible genome complement of the commensal *Escherichia coli* strain A0 34/86 (O83:K24:H31). *Microbiology* 151 (pt. 2):385–98.

Herring, C. D., A. Raghunathan, C. Honisch, T. Patel, M. K. Applebee, A. R. Joyce, T. J. Albert, F. R. Blattner, D. van den Boom, C. R. Cantor, and B. O. Palsson. 2006. Comparative genome sequencing of *Escherichia coli* allows observation of bacterial evolution on a laboratory timescale. *Nat Genet* 38 (12):1406–12.

Hershey, A. D., and M. Chase. 1952. Independent functions of viral protein and nucleic acid in growth of bacteriophage. *J Gen Physiol* 36 (1):39–56.

Higgins, Norman Patrick. 2005. *The bacterial chromosome.* Washington, D.C.: ASM Press.

Holmes, Frederick Lawrence. 2001. *Meselson, Stahl, and the replication of DNA: A history of the "most beautiful experiment in biology."* New Haven: Yale University Press.

Hou, Shaobin, Jimmy H. Saw, Kit Shan Lee, Tracey A. Freitas, Claude Belisle, Yutaka Kawarabayasi, Stuart P. Donachie, Alla Pikina, Michael Y. Galperin, Eugene V. Koonin, Kira S. Makarova, Marina V. Omelchenko, Alexander Sorokin, Yuri I. Wolf, Qing X. Li, Young Soo Keum, Sonia Campbell, Judith Denery, Shin-Ichi Aizawa, Satoshi Shibata, Alexander Malahoff, and Maqsudul Alam. 2004. Genome sequence of the deep-sea γ-proteobacterium *Idiomarina loihiensis* reveals amino acid fermentation as a source of carbon and energy. *Proc Natl Acad Sci USA* 101 (52):18036–41.

Humes, Edward. 2007. *Monkey girl: Evolution, education, religion, and the battle for America's soul.* New York: Ecco.

Inoue, T., R. Shingaki, S. Hirose, K. Waki, H. Mori, and K. Fukui. 2007. Genome-wide screening of genes required for swarming motility in *Escherichia coli* K-12. *J Bacteriol* 189(3):950–57.

Jackson, David Archer, and Stephen P. Stich. 1979. *The recombinant DNA debate.* Englewood Cliffs, N.J.: Prentice-Hall.

Jacob, François. 1995. *The statue within: An autobiography.* Cold Spring Harbor, N.Y.: Cold Spring Harbor Laboratory Press.

Jennison, A. V., and N. K. Verma. 2004. *Shigella flexneri* infection: Pathogenesis and vaccine development. *FEMS Microbiol Rev* 28 (1):43–58.

Jones, Nancy. 2006. *Round three—"mixing and matching" biological building blocks: Mouse-human chimeras are here!* Center for Bioethics and Human Dignity. http://www.cbhd.org/resources/biotech/jones_2006-01-20.htm.

Jones, Shari A., F. A. Chowdhury, et al. 2007. Respiration of *Escherichia coli* in the mouse intestine. *Infect Immun* 75:4891–99.

Judson, Horace Freeland. 1996. *The eighth day of creation: Makers of the revolution in biology.* Expanded ed. Cold Spring Harbor, N.Y.: Cold Spring Harbor Laboratory Press.

Jun, S., and B. Mulder. 2006. Entropy-driven spatial organization of highly confined polymers: Lessons for the bacterial chromosome. *Proc Natl Acad Sci USA* 103 (33):12388–93.

Kalir, S., S. Mangan, and U. Alon. 2005. A coherent feed-forward loop with a SUM input function prolongs flagella expression in *Escherichia coli. Mol Syst Biol* 1:2005.0006.

Kaper, J. B. 2005. Pathogenic *Escherichia coli. Int J Med Microbiol* 295 (6–7):355–56.

Karch, H., P. I. Tarr, and M. Bielaszewska. 2005. Enterohaemorrhagic *Escherichia coli* in human medicine. *Int J Med Microbiol* 295 (6–7):405–18.

Karp, P. D., I. M. Kessler, A. Shearer, et al. 2007. Multidimensional annotation of the *Escherichia coli.* K-12 genome. *Nucleic Acids Res* 35:7577–90.

Kass, Leon. 1997. National Bioethics Advisory Commission. Testimony of March 14. CyberCemetery. University of North Texas Libraries. http://govinfo.library.unt.edu/nbac/transcripts/1997/3-14-97.pdf.

Kass, Leon, and James Q. Wilson. 1998. *The ethics of human cloning.* Washington, D.C.: AEI Press.

Kerr, B., M. A. Riley, M. W. Feldman, and B. J. Bohannan. 2002. Local dispersal promotes biodiversity in a real-life game of rock-paper-scissors. *Nature* 418 (6894):171–74.

Kirkup, B. C., and M. A. Riley. 2004. Antibiotic-mediated antagonism leads to a bacterial game of rock-paper-scissors in vivo. *Nature* 428 (6981):412–14.

Kobayashi, I. 2001. Behavior of restriction-modification systems as selfish mobile elements and their impact on genome evolution. *Nucleic Acids Res* 29 (18):3742–56.

Krawinkel, M. B. 2007. What we know and don't know about Golden Rice. *Nat Biotechnol* 25 (6):623.

Krimsky, Sheldon. 1982. *Genetic alchemy: The social history of the recombinant DNA controversy.* Cambridge, Mass.: MIT Press.

Kropotkin, Petr Alekseevich. 1919. *Mutual aid: A factor of evolution.* New York: Alfred A. Knopf.

Kurata, H., H. El-Samad, R. Iwasaki, H. Ohtake, J. C. Doyle, I. Grigorova, C. A. Gross, and M. Khammash. 2006. Module-based analysis of robustness tradeoffs in the heat shock response system. *PLoS Comput Biol* 2 (7):e59.

Kussell, E., R. Kishony, N. Q. Balaban, and S. Leibler. 2005. Bacterial persistence: A model of survival in changing environments. *Genetics* 169 (4):1807–14.

La Duc, M. T., K. Venkateswaran, R. Sumner, and D. Pierson. 2004. Characterization and monitoring of microbes in the international space station drinking water. Beacon eSpace. Jet Propulsion Laboratory. http://hdl.handle.net/2014/37245.

Lamarck, Jean Baptiste Pierre Antoine de Monet de. 1984. *Zoological philosophy: An exposition with regard to the natural history of animals.* Chicago: University of Chicago Press.

Lambert, C., K. J. Evans, R. Till, L. Hobley, M. Capeness, S. Rendulic, S. C. Schuster, S. Aizawa, and R. E. Sockett. 2006. Characterizing the flagellar filament and the role of motility in bacterial prey-penetration by *Bdellovibrio bacteriovorus. Mol Microbiol* 60 (2):274–86.

Lawrence, J. G., and H. Ochman. 1998. Molecular archaeology of the *Escherichia coli* genome. *Proc Natl Acad Sci USA* 95 (16):9413–17.

Lawrence, S. 2007a. Agbiotech booms in emerging nations. *Nat Biotechnol* 25 (3):271.

———. 2007b. State of the biotech sector—2006. *Nat Biotechol* 25:706.

Lederberg, J. 1946. Bacterial genetics. Joshua Lederberg: Biomedical Science and the Public Interest. National Library of Medicine. http://profiles.nlm.nih.gov/hmd/lederberg/bacterial/html.

———. 1960. Exobiology: Approaches to life beyond the earth. *Science* 132 (3424):393–400.

———. 1963. Life beyond Earth. *Stanford Today,* Winter.

———. 1987. Genetic recombination in bacteria: A discovery account. *Annu Rev Genet* 21:23–46.

Lederberg, J., and E. M. Lederberg. 1952. Replica plating and indirect selection of bacterial mutants. *J Bacteriol* 63 (3):399–406.

Lenski, R. 2003. The ecology, genetics and evolution of bacteria in an experimental setting. *Curr Biol* 13 (12):R466–67.

Lerat, E., V. Daubin, H. Ochman, and N. A. Moran. 2005. Evolutionary origins of genomic repertoires in bacteria. *PLoS Biol* 3 (5):e130.

Levskaya, A., A. A. Chevalier, J. J. Tabor, Z. B. Simpson, L. A. Lavery, M. Levy, E. A. Davidson, A. Scouras, A. D. Ellington, E. M. Marcotte, and C. A. Voigt. 2005. Synthetic biology: Engineering *Escherichia coli* to see light. *Nature* 438 (7067):441–42.

Levy, Stuart B. 2002. *The antibiotic paradox: How the misuse of antibiotics destroys their curative powers.* 2nd ed. Cambridge, Mass.: Perseus.

Lewis, K. 2005. Persister cells and the riddle of biofilm survival. *Biochemistry (Mosc)* 70 (2):267–74.

———. 2007. Persister cells, dormancy, and infectious disease. *Nat Rev Microbiol* 5 (1):48–56.

Lim, H. N., and A. van Oudenaarden. 2007. A multistep epigenetic switch enables the stable inheritance of DNA methylation states. *Nat Genet* 39 (2):269–75.

Lumsden, Richard D. 1994. Not so blind a watchmaker. *Creation Research Society Quarterly Journal* 31 (1).

Luria, S. E. 1984. *A slot machine, a broken test tube: An autobiography.* Alfred P. Sloan Foundation Series. New York: Harper & Row.

Luria, S. E., and M. Delbrück. 1943. Mutations of bacteria from virus sensitivity to virus resistance. *Genetics* 28(6):491–511.

Luria, S. E., M. Delbrück, and T. F. Anderson. 1943. Electron microscope studies of bacterial viruses. *J Bacteriol* 46 (1):57–77.

Ma, H. W., and A. P. Zeng. 2003. The connectivity structure, giant strong component and centrality of metabolic networks. *Bioinformatics* 19 (11):1423–30.

Margolin, W. 2005. FtsZ and the division of prokaryotic cells and organelles. *Nat Rev Mol Cell Biol* 6 (11):862–71.

Marshall, R. E., C. T. Caskey, and M. Nirenberg. 1967. Fine structure of RNA codewords recognized by bacterial, amphibian, and mammalian transfer RNA. *Science* 155 (764):820–26.

Matz, C., and S. Kjelleberg. 2005. Off the hook—how bacteria survive protozoan grazing. *Trends Microbiol* 13 (7):302–7.

Matz, C., D. McDougald, A. M. Moreno, P. Y. Yung, F. H. Yildiz, and S. Kjelleberg. 2005. Biofilm formation and phenotypic variation enhance predation-driven persistence of *Vibrio cholerae. Proc Natl Acad Sci USA* 102 (46):16819–24.

McElheny, Victor. 1976. Hunt for evidence of life on Mars is still a puzzle. *New York Times,* August 11, 1976.

Meltz Steinberg, K., and B. R. Levin. 2007. Grazing protozoa and the evolution of the *Escherichia coli* O157:H7 Shiga toxin–encoding prophage. *Proc Biol Sci* 274 (1621):1924–29.

Meselson, M., and F. W. Stahl. 1958. The replication of DNA in *Escherichia coli. Proc Nat Acad Sci USA* 44:671–82.

Mettetal, J. T., D. Muzzey, J. M. Pedraza, E. M. Ozbudak, and A. van Oudenaarden. 2006. Predicting stochastic gene expression dynamics in single cells. *Proc Natl Acad Sci USA* 103 (19):7304–9.

Michaelian, K. 2005. Thermodynamic stability of ecosystems. *J Theor Biol* 237 (3):323–35.

Michod, R. E. 2007. Evolution of individuality during the transition from unicellular to multicellular life. *Proc Natl Acad Sci USA* 104 (1):S8613–18.

Milo, R., S. Shen-Orr, S. Itzkovitz, N. Kashtan, D. Chklovskii, and U. Alon. 2002. Network motifs: Simple building blocks of complex networks. *Science* 298 (5594):824–27.

Mira, A., R. Pushker, and F. Rodriguez-Valera. 2006. The Neolithic revolution of bacterial genomes. *Trends Microbiol* 14 (5):200–206.

Morbey, Y. E., C. E. Brassil, and A. P. Hendry. 2005. Rapid senescence in Pacific salmon. *Am Nat* 166 (5):556–68.

Mrak, P., Z. Podlesek, J. P. van Putten, and D. Zgur-Bertok. 2007. Heterogeneity in expression of the *Escherichia coli* colicin K activity gene cka is controlled by the SOS system and stochastic factors. *Mol Genet Genomics* 277 (4):391–401.

Mulec, J., Z. Podlesek, P. Mrak, A. Kopitar, A. Ihan, and D. Zgur-Bertok. 2003. A cka-gfp transcriptional fusion reveals that the colicin K activity gene is induced in only 3 percent of the population. *J Bacteriol* 185 (2):654–59.

Müller-Hill, Benno. 1996. *The* lac *operon: A short history of a genetic paradigm.* New York: Walter de Gruyter.

Myllykangas, S., J. Himberg, T. Böhling, B. Nagy, J. Hollmén, and S. Knuutila. 2006. DNA copy number amplification profiling of human neoplasms. *Oncogene* 25 (55):7324–32.

Napier, W. M. 2004. A mechanism for interstellar panspermia. *Monthly Notices of the Royal Astronomical Society* 348 (1):46–51.

Nash, J. M. 2000. Grains of hope. *Time,* July 31, 2000, 38–46.

National Academy of Sciences. 1977. *Research with recombinant DNA: An Academy Forum, March 7–9, 1977.* Washington, D.C.: National Academy of Sciences.

Naylor, S. W., D. L. Gally, and J. C. Low. 2005. Enterohaemorrhagic *E. coli* in veterinary medicine. *Int J Med Microbiol* 295 (6–7):419–41.

Neidhardt, Frederick C. 1996. Escherichia coli *and* Salmonella: *Cellular and molecular biology.* 2nd ed. Washington, D.C.: ASM Press.

Nicholson, J. K., E. Holmes, and I. D. Wilson. 2005. Gut microorganisms, mammalian metabolism and personalized health care. *Nat Rev Microbiol* 3 (5):431–38.

Nielsen, K. M., and J. P. Townsend. 2004. Monitoring and modeling horizontal gene transfer. *Nat Biotechnol* 22 (9):1110–14.

Noireaux, V., R. Bar-Ziv, J. Godefroy, H. Salman, and A. Libchaber. 2005. Toward an artificial cell based on gene expression in vesicles. *Phys Biol* 2 (3):P1–8.

Nordenberg, T. 1999. Maker of growth hormone feels long arm of law. *FDA Consum* 33 (5):33.

Norris, V., T. den Blaauwen, A. Cabin-Flaman, R. H. Doi, R. Harshey, L. Janniere, A. Jimenez-Sanchez, D. J. Jin, P. A. Levin, E. Mileykovskaya, A. Minsky, M. Saier Jr., and K. Skarstad. 2007. Functional taxonomy of bacterial hyperstructures. *Microbiol Mol Biol Rev* 71 (1):230–53.

Novick, A., and M. Weiner. 1957. Enzyme induction as an all-or-none phenomenon. *Proc Natl Acad Sci USA* 43 (7):553–66.

Nystrom, T. 2004. Stationary-phase physiology. *Annu Rev Microbiol* 58:161–81.

O'Donnell, M. 2006. Replisome architecture and dynamics in *Escherichia coli. J Biol Chem* 281 (16):10653–56.

Ostrowski, E. A., D. E. Rozen, and R. E. Lenski. 2005. Pleiotropic effects of beneficial mutations in *Escherichia coli. Evolution Int J Org Evolution* 59 (11):2343–52.

Ouzounis, C. A., V. Kunin, N. Darzentas, and L. Goldovsky. 2006. A minimal estimate for the gene content of the last universal common ancestor—exobiology from a terrestrial perspective. *Res Microbiol* 157 (1):57–68.

Owen, M. D., and I. A. Zelaya. 2005. Herbicide-resistant crops and weed resistance to herbicides. *Pest Manag Sci* 61 (3):301–11.

Ozbudak, E. M., M. Thattai, H. N. Lim, B. I. Shraiman, and A. Van Oudenaarden. 2004. Multistability in the lactose utilization network of *Escherichia coli. Nature* 427 (6976):737–40.

Pallen, M. J., and N. J. Matzke. 2006. From *The Origin of Species* to the origin of bacterial flagella. *Nat Rev Microbiol* 4 (10):784–90.

Park, H. K., S. S. Shim, S. Y. Kim, J. H. Park, S. E. Park, H. J. Kim, B. C. Kang, and C. M. Kim. 2005. Molecular analysis of colonized bacteria in a human newborn infant gut. *J Microbiol* 43 (4):345–53.

Park, S., P. M. Wolanin, E. A. Yuzbashyan, P. Silberzan, J. B. Stock, and R. H. Austin. 2003. Motion to form a quorum. *Science* 301 (5630):188.

Patrick, W. M., E. M. Quandt, D. V. Swartzlander, and I. Matsumura. 2007. Multicopy suppression underpins metabolic evolvability. *Mol Bio Evo* 24 (12):2716–22.

Pelosi, L., L. Kuhn, D. Guetta, J. Garin, J. Geiselmann, R. E. Lenski, and D. Schneider. 2006. Parallel changes in global protein profiles during long-term experimental evolution in *Escherichia coli. Genetics* 173 (4):1851–69.

Pennington, T. H. 2003. *When food kills: BSE,* E. coli, *and disaster science.* New York: Oxford University Press.

Perfeito, L., L. Fernandes, C. Mota, and I. Gordo. 2007. Adaptive mutations in bacteria: High rate and small effects. *Science* 317:813–15.

Perna, N. T., G. Plunkett III, V. Burland, B. Mau, J. D. Glasner, D. J. Rose, G. F. Mayhew, P. S. Evans, J. Gregor, H. A. Kirkpatrick, G. Posfai, J. Hackett, S. Klink, A. Boutin, Y. Shao, L. Miller, E. J. Grotbeck, N. W. Davis, A. Lim, E. T. Dimalanta, K. D. Potamousis, J. Apodaca, T. S. Anantharaman, J. Lin, G. Yen, D. C. Schwartz, R. A. Welch, and F. R. Blattner. 2001. Genome sequence of enterohaemorrhagic *Escherichia coli* O157:H7. *Nature* 409 (6819):529–33.

Perron, G. G., M. Zasloff, and G. Bell. 2006. Experimental evolution of resistance to an antimicrobial peptide. *Proc Biol Sci* 273 (1583):251–56.

Peschel, A., and H. G. Sahl. 2006. The co-evolution of host cationic antimicrobial peptides and microbial resistance. *Nat Rev Microbiol* 4 (7):529–36.

Ponder, R. G., N. C. Fonville, and S. M. Rosenberg. 2005. A switch from high-fidelity to error-prone DNA double-strand break repair underlies stress-induced mutation. *Mol Cell* 19 (6):791–804.

Potrykus, I. 2001. Golden Rice and beyond. *Plant Physiol* 125 (3):1157–61.

Power, M. L., J. Littlefield-Wyer, D. M. Gordon, D. A. Veal, and M. B. Slade. 2005. Phenotypic and genotypic characterization of encapsulated *Escherichia coli* isolated from blooms in two Australian lakes. *Environ Microbiol* 7 (5):631–40.

Queitsch, C., T. A. Sangster, and S. Lindquist. 2002. Hsp90 as a capacitor of phenotypic variation. *Nature* 417 (6889):618–24.

Raney, T. 2006. Economic impact of transgenic crops in developing countries. *Curr Opin Biotechnol* 17 (2):174–78.

Rangel, J. M., P. H. Sparling, C. Crowe, P. M. Griffin, and D. L. Swerdlow. 2005. Epidemiology of *Escherichia coli* O157:H7 outbreaks, United States, 1982–2002. *Emerg Infect Dis* 11 (4):603–9.

Raser, J. M., and E. K. O'Shea. 2005. Noise in gene expression: Origins, consequences, and control. *Science* 309 (5743):2010–13.

Rasheed, J. K., C. Jay, B. Metchock, F. Berkowitz, L. Weigel, J. Crellin, C. Steward, B. Hill, A. A. Medeiros, and F. C. Tenover. 1997. Evolution of extended-spectrum beta-lactam resistance (SHV-8) in a strain of *Escherichia coli* during multiple episodes of bacteremia. *Antimicrob Agents Chemother* 41 (3):647–53.

Raymond, J., and D. Segre. 2006. The effect of oxygen on biochemical networks and the evolution of complex life. *Science* 311 (5768):1764–67.

Reinhold, Robert. 1969. Scientists isolate a gene: Steps in heredity control. *New York Times,* November 22, 1969.

Reisner, A., K. A. Krogfelt, B. M. Klein, E. L. Zechner, and S. Molin. 2006. In vitro biofilm formation of commensal and pathogenic *Escherichia coli* strains: Impact of environmental and genetic factors. *J Bacteriol* 188 (10):3572–81.

Remold, S. K., and R. E. Lenski. 2004. Pervasive joint influence of epistasis and plasticity on mutational effects in *Escherichia coli. Nat Genet* 36 (4):423–26.

Ren, C. P., S. A. Beatson, J. Parkhill, and M. J. Pallen. 2005. The Flag-2 locus, an ancestral gene cluster, is potentially associated with a novel flagellar system from *Escherichia coli. J Bacteriol* 187 (4):1430–40.

Roberts, Richard B. 1955. *Studies of biosynthesis in* Escherichia coli. Washington, D.C.: Carnegie Institution.

Robicsek, A., J. Strahilevitz, G. A. Jacoby, M. Macielag, D. Abbanat, C. H. Park, K. Bush, and D. C. Hooper. 2006. Fluoroquinolone-modifying enzyme: A new adaptation of a common aminoglycoside acetyltransferase. *Nat Med* 12 (1):83–88.

Rogers, Michael. 1977. *Biohazard.* New York: Alfred A. Knopf.

Roth, J. R., E. Kugelberg, A. B. Reams, E. Kofoid, and D. I. Andersson. 2006. Origin of mutations under selection: The adaptive mutation controversy. *Annu Rev Microbiol* 60:477–501.

Rozen, D. E., D. Schneider, and R. E. Lenski. 2005. Long-term experimental evolution in *Escherichia coli.* XIII. Phylogenetic history of a balanced polymorphism. *J Mol Evol* 61 (2):171–80.

Salyers, Abigail A., and Dixie D. Whitt. 2005. *Revenge of the microbes: How bacterial resistance is undermining the antibiotic miracle.* Washington, D.C.: ASM Press.

Sandermann, H. 2006. Plant biotechnology: Ecological case studies on herbicide resistance. *Trends Plant Sci* 11 (7):324–28.

Sansonetti, P. J. 2006. Shigellosis: An old disease in new clothes? *PLoS Med* 3 (9):e354.

Sapp, J. 1994. *Evolution by association: A history of symbiosis.* New York: Oxford University Press.

———. 2005. The prokaryote-eukaryote dichotomy: Meanings and mythology. *Microbiol Mol Biol Rev* 69 (2):292–305.

Sauer, U. 2006. Metabolic networks in motion: ^{13}C-based flux analysis. *Mol Syst Biol* 2:62.

Sauer, U., M. Heinemann, and N. Zamboni. 2007. Genetics: Getting closer to the whole picture. *Science* 316 (5824):550–51.

Schaechter, Moselio, John L. Ingraham, and Frederick C. Neidhardt. 2006. *Microbe.* Washington, D.C.: ASM Press.

Schnapp, N., and Q. Schiermeier. 2001. Critics claim "sight-saving" rice is over-rated. *Nature* 410 (6828):503.

Schneider, D., and R. E. Lenski. 2004. Dynamics of insertion sequence elements during experimental evolution of bacteria. *Res Microbiol* 155 (5):319–27.

Scott, C. T. 2006. Chimeras in the crosshairs. *Nat Biotechnol* 24 (5):487–90.

Segerstrale, Ullica. In press. *Nature's oracle: A life of W. D. Hamilton.* Oxford: Oxford University Press.

Service, Robert F. 2003. Researchers create first autonomous synthetic life form. *Science* 299 (5607):640.

———. 2007. A growing threat down on the farm. *Science* 316 (5828):1114–17.

Shaikh, N., and P. I. Tarr. 2003. *Escherichia coli* O157:H7 Shiga toxin–encoding bacteriophages: Integrations, excisions, truncations, and evolutionary implications. *J Bacteriol* 185 (12):3596–605.

Shapiro, J., L. Machattie, L. Eron, G. Ihler, K. Ippen, and J. Beckwith. 1969. Isolation of pure *lac* operon DNA. *Nature* 224 (5221):768–74.

Shi, W., and D. R. Zusman. 1993. Fatal attraction. *Nature* 366 (6454):414–15.

Shin, T., D. Kraemer, J. Pryor, L. Liu, J. Rugila, L. Howe, S. Buck, K. Murphy, L. Lyons, and M. Westhusin. 2002. A cat cloned by nuclear transplantation. *Nature* 415 (6874):859.

Shnayerson, Michael, and Mark J. Plotkin. 2002. *The killers within: The deadly rise of drug-resistant bacteria.* Boston: Little, Brown.

Silver, Lee M. 2006. *Challenging nature: The clash of science and spirituality at the frontiers of life.* New York: Ecco.

Sinervo, B. 2001. Runaway social games, genetic cycles driven by alternative male and female strategies, and the origin of morphs. *Genetica* 112–13:417–34.

Singer, M. 2001. What did the Asilomar exercise accomplish, what did it leave undone? *Perspect Biol Med* 44 (2):186–91.

Spencer, C. C., M. Bertrand, M. Travisano, and M. Doebeli. 2007. Adaptive diversification in genes that regulate resource use in *Escherichia coli. PLoS Genet* 3 (1):e15.

Spudich, J. L., and D. E. Koshland Jr. 1976. Non-genetic individuality: Chance in the single cell. *Nature* 262 (5568):467–71.

Stahl, Franklin W. 2001. Alfred Day Hershey. Biographical Memoirs 80. Washington, D.C.: National Academy Press.

Stein, A. J., H. P. Sachdev, and M. Qaim. 2006. Potential impact and cost-effectiveness of Golden Rice. *Nat Biotechnol* 24 (10):1200–1201.

Stenuit, B., L. Eyers, R. Rozenberg, J. L. Habib-Jiwan, and S. N. Agathos. 2006. Aerobic growth of *Escherichia coli* with 2,4,6-trinitrotoluene (TNT) as the sole nitrogen source and evidence of TNT denitration by whole cells and cell-free extracts. *Appl Environ Microbiol* 72 (12):7945–48.

Stewart, E. J., R. Madden, G. Paul, and F. Taddei. 2005. Aging and death in an organism that reproduces by morphologically symmetric division. *PLoS Biol* 3 (2):e45.

Sullivan, Woodruff T., and John Baross, eds. 2007. *Planets and life: The emerging science of astrobiology.* New York: Cambridge University Press.

Talbot, M. 2005. Darwin in the dock. *New Yorker,* December 5, 2005.

TalkOrigins Archive. 2006. *Kitzmiller v. Dover Area School District,* Dover, Pennsylvania Intelligent Design Case. http://talkorigins.org/faqs/dover/kitzmiller_v_dover.html.

Tanaka, R., M. Csete, and J. Doyle. 2005. Highly optimised global organisation of metabolic networks. *Syst Biol (Stevenage)* 152 (4):179–84.

Tarr, P. I., C. A. Gordon, and W. L. Chandler. 2005. Shiga-toxin-producing *Escherichia coli* and haemolytic uraemic syndrome. *Lancet* 365 (9464):1073–86.

Tatum, E. L., and J. Lederberg. 1947. Gene recombination in the bacterium *Escherichia coli. J Bacteriol* 53 (6):673–84.

Tenaillon, O., E. Denamur, and I. Matic. 2004. Evolutionary significance of stress-induced mutagenesis in bacteria. *Trends Microbiol* 12 (6):264–70.

Tenover, F. C. 2006. Mechanisms of antimicrobial resistance in bacteria. *Am J Med* 119 (6 Suppl. 1):S3–10, discussion S62–70.

Thanbichler, M., and L. Shapiro. 2006. Chromosome organization and segregation in bacteria. *J Struct Biol* 156 (2):292–303.

Thiem, S., D. Kentner, and V. Sourjik. 2007. Positioning of chemosensory clusters in *E. coli* and its relation to cell division. *Embo J* 26 (6):1615–23.

Thomas, W. E., L. M. Nilsson, M. Forero, E. V. Sokurenko, and V. Vogel. 2004. Shear-dependent "stick-and-roll" adhesion of type 1 fimbriated *Escherichia coli. Mol Microbiol* 53 (5):1545–57.

Thornton, I.W.B. 1996. *Krakatau: The destruction and reassembly of an island ecosystem.* Cambridge, Mass.: Harvard University Press.

Tomitani, A., A. H. Knoll, C. M. Cavanaugh, and T. Ohno. 2006. The evolutionary diversification of cyanobacteria: Molecular-phylogenetic and paleontological perspectives. *Proc Natl Acad Sci USA* 103 (14):5442–47.

Travisano, M., J. A. Mongold, A. F. Bennett, and R. E. Lenski. 1995. Experimental tests of the roles of adaptation, chance, and history in evolution. *Science* 267 (5194):87–90.

Trinh, C. T., R. Carlson, A. Wlaschin, and F. Srienc. 2006. Design, construction and performance of the most efficient biomass-producing *E. coli* bacterium. *Metab Eng* 8 (6):628–38.

U.S. Congress. Senate. 2005. *Human Chimera Prohibition Act of 2005.* S 659. 109th Cong., 1st sess.

U.S. Department of Health and Human Services, Centers for Disease Control and Prevention. 2006. Multi-state outbreak of *E. coli* O157:H7 infections from spinach. September–October. http://www.cdc.gov/ecoli2006/September/.

Van Till, Howard J. 2002. *E. coli* at the *No Free Lunchroom:* Bacterial flagella and Dembski's case for intelligent design. American Association for the Advancement of Science. http://www.aaas.org/spp/dser/03_Areas/evolution/perspectives/vantillecoli_2002.pdf.

Varma, J. K., K. D. Greene, M. E. Reller, S. M. DeLong, J. Trottier, S. F. Nowicki, M. DiOrio, E. M. Koch, T. L. Bannerman, S. T. York, M. A. Lambert-Fair, J. G. Wells, and P. S. Mead. 2003. An outbreak of *Escherichia coli* O157 infection following exposure to a contaminated building. *JAMA* 290 (20):2709–12.

Vulić, M., and R. Kolter. 2001. Evolutionary cheating in *Escherichia coli* stationary phase cultures. *Genetics* 158 (2):519–26.

Wade, Nicholas. 1977. *The ultimate experiment: Man-made evolution.* New York: Walker.

Walters, M., and V. Sperandio. 2006. Quorum sensing in *Escherichia coli* and *Salmonella. Int J Med Microbiol* 296 (2–3):125–31.

Wandersman, C., and P. Delepelaire. 2004. Bacterial iron sources: From siderophores to hemophores. *Annu Rev Microbiol* 58:611–47.

Wang, L., J. Xie, and P. G. Schultz. 2006. Expanding the genetic code. *Annu Rev Biophys Biomol Struct* 35:225–49.

Warmflash, D., and B. Weiss. 2005. Did life come from another world? *Sci Am* 293 (5):64–71.

Watanabe, T. 1963. Infective heredity of multiple drug resistance in bacteria. *Bacteriol Rev* 27:87–115.

Watson, James D. 1969. *The double helix: A personal account of the discovery of the structure of DNA.* New York: New American Library.

Watson, James D., and John Tooze. 1981. *The DNA story: A documentary history of gene cloning.* San Francisco: W. H. Freeman.

Welch, R. A., V. Burland, G. Plunkett III, P. Redford, P. Roesch, D. Rasko, E. L. Buckles, S. R. Liou, A. Boutin, J. Hackett, D. Stroud, G. F. Mayhew, D. J. Rose, S. Zhou, D. C. Schwartz, N. T. Perna, H. L. Mobley, M. S. Donnenberg, and F. R. Blattner. 2002. Extensive mosaic structure revealed by the complete genome sequence of uropathogenic *Escherichia coli. Proc Natl Acad Sci USA* 99 (26):17020–24.

Wells, H. G. 1896. *The island of Dr. Moreau.* Garden City, N.Y.: Garden City Publishing.

West, S. A., A. S. Griffin, A. Gardner, and S. P. Diggle. 2006. Social evolution theory for microorganisms. *Nat Rev Microbiol* 4 (8):597–607.

White-Ziegler, C. A., A. J. Malhowski, and S. Young. 2007. Human body temperature (37 C) increases the expression of iron, carbohydrate, and amino acid utilization genes in *Escherichia coli* K-12. *J Bacteriol* 189:5429–40.

Wick, L. M., W. Qi, D. W. Lacher, and T. S. Whittam. 2005. Evolution of genomic content in the stepwise emergence of *Escherichia coli* O157:H7. *J Bacteriol* 187 (5):1783–91.

Willenbrock, H., and D. W. Ussery. 2004. Chromatin architecture and gene expression in *Escherichia coli. Genome Biol* 5 (12):252.

Williams, G. C. 1966. *Adaptation and natural selection.* Princeton: Princeton University Press.

———. 1999. The 1999 Crafoord Prize lectures. The Tithonus error in modern gerontology. *Q Rev Biol* 74 (4):405–15.

Wirth, T., D. Falush, R. Lan, F. Colles, P. Mensa, L. H. Wieler, H. Karch, P. R. Reeves, M. C. Maiden, H. Ochman, and M. Achtman. 2006. Sex and virulence in *Escherichia coli:* An evolutionary perspective. *Mol Microbiol* 60 (5):1136–51.

Wise, R., and E. J. Soulsby. 2002. Antibiotic resistance—an evolving problem. *Vet Rec* 151 (13):371–72.

Woese, C. R., and G. E. Fox. 1977. Phylogenetic structure of the prokaryotic domain: The primary kingdoms. *Proc Natl Acad Sci USA* 74 (11):5088–90.

Woldringh, C. L., and N. Nanninga. 2006. Structural and physical aspects of bacterial chromosome segregation. *J Struct Biol* 156 (2):273–83.

Wolf, D. M., V. V. Vazirani, and A. P. Arkin. 2005. Diversity in times of adversity: Probabilistic strategies in microbial survival games. *J Theor Biol* 234 (2):227–53.

Wolfe, A. J. 2005. The acetate switch. *Microbiol Mol Biol Rev* 69 (1):12–50.

Woods, R., D. Schneider, C. L. Winkworth, M. A. Riley, and R. E. Lenski. 2006. Tests of parallel molecular evolution in a long-term experiment with *Escherichia coli*. *Proc Natl Acad Sci USA* 103 (24):9107–12.

Wright, Susan. 1994. *Molecular politics: Developing American and British regulatory policy for genetic engineering, 1972–1982.* Chicago: University of Chicago Press.

Xavier, J. B., and K. R. Foster. 2007. Cooperation and conflict in microbial biofilms. *Proc Natl Acad Sci USA* 104 (3):876–81.

Zelaya, Ian A., Michael D. K. Owen, and Mark J. VanGessel. 2007. Transfer of glyphosate resistance: Evidence of hybridization in *Conyza* (Asteraceae). *Am J Bot* 94 (4):660–73.

Zhang, W., W. Qi, T. J. Albert, A. S. Motiwala, D. Alland, E. K. Hyytia-Trees, E. M. Ribot, P. I. Fields, T. S. Whittam, and B. Swaminathan. 2006. Probing genomic diversity and evolution of *Escherichia coli* O157 by single nucleotide polymorphisms. *Genome Res* 16 (6):757–67.

Zhou, T., J. M. Carlson, and J. Doyle. 2005. Evolutionary dynamics and highly optimized tolerance. *J Theor Biol* 236 (4):438–47.

Zilinskas, Raymond A., and Burke K. Zimmerman. 1986. *The gene-splicing wars: Reflections on the recombinant DNA controversy.* New York: Macmillan.

Zimmer, C. 2004. George C. Williams profile: Stretching the limits of evolutionary biology. *Science* 304 (5675):1235–36.

———. 2006. Did DNA come from viruses? *Science* 312 (5775):870–72.

———. 2007a. Evolved for cancer? *Sci Am* 296 (1):68–74, 75A.

———. 2007b. Fast-reproducing microbes provide a window on natural selection. *New York Times*, June 26, 2007.

Zinser, E. R., and R. Kolter. 2004. *Escherichia coli* evolution during stationary phase. *Res Microbiol* 155 (5):328–36.

Zorzano, M. P., D. Hochberg, M. T. Cuevas, and J. M. Gómez-Gómez. 2005. Reaction-diffusion model for pattern formation in *E. coli* swarming colonies with slime. *Phys Rev E Stat Nonlin Soft Matter Phys* 71 (3 pt. 1):031908.

INDEX

Page numbers in *italics* refer to illustrations.

A NOTE ON THE TYPE

This book was set in Minion, a typeface produced by the Adobe Corporation specifically for the Macintosh personal computer and released in 1990. Designed by Robert Slimbach, Minion combines the classic characteristics of old-style faces with the full complement of weights required for modern typesetting.

Composed by North Market Street Graphics, Lancaster, Pennsylvania

Printed and bound by Berryville Graphics, Berryville, Virginia

Designed by M. Kristen Bearse